Menas C. Kafatos

SCIENCE, REALITY & EVERYDAY LIFE

Science, Reality & Everyday Life
Menas C. Kafatos

First edition 2024

Copyright credits:
Pink Integer Books,
1275 Pine Creek Way #B
Concord, CA 94520, USA.
pinkinteger@indiansacredculture.org

Print ISBN: 978-81-988068-1-9
eBook ISBN: 978-81-988068-6-4

BISAC Codes
SCI057000   SCIENCE / Physics / Quantum Theory
REL032000   RELIGION / Hinduism / General
REL032010   RELIGION / Hinduism / History
SCI000000   SCIENCE / General
SCI019000   SCIENCE / Earth Sciences / General
SCI024000   SCIENCE / Energy
OCC000000   BODY, MIND & SPIRIT / General
OCC027000   BODY, MIND & SPIRIT / Spiritualism
PHI000000   PHILOSOPHY / General PHILOSOPHY / Philosophy of
Science see SCIENCE / Philosophy & Social Aspects

Thema Subject Categories
PHQ   Quantum physics (quantum mechanics and quantum field theory)
QDHC   East Asian and Indian philosophy
PS   Biology, life sciences
VX   Mind, body, spirit
VXK   Earth energies

Published by:
PRISMA, an imprint of
Digital Media Initiatives | DMI Systems Pvt Ltd
PRISMA, Aurelec/ Prayogshala,
Auroville 605101, Tamil Nadu, India
www.prisma.haus

*Dedicated to the sincere students of
Consciousness*

Foreword

What is this book that fell into your hands? It is an exploration of Reality, which examines Truth scientifically, philosophically (mainly from the point of view of the Triadic, non-dual system of understanding) which originated in ancient India with connections to Greek philosophical systems, Buddhism and Taoism. We look at parallels to 'Western teachings of the philosophical schools and the great teachers, among them Heraclitus, Socrates, Plato, Aristotle, the Neoplatonists and the Stoics. We see some parallels to the idea of Trinity encountered in religion. However, the Triadic System is not a religion. It does not deny religion, it does not replace religion. In the same way, it does not replace science. On the contrary, the Triadic System helps us to uncover the deeper meaning of religion, if we are believing, and the deeper meaning of science, if we are searching and exploring. But most importantly, it may help us to see that these two "poles" are not opposites, they are complementary to each other. Of course, the present treatise does not go against any philosophy, ancient or modern, because it holds that after all, despite all the apparent dualities, they all originate from the same universal Source.

This book is both a theoretical and a practical handbook. Many of the truths it expresses have to do with everyday life, approaching life in a scientific way. However, don't expect to "learn" the truth by only reading. Certainly what is written here is concise but not simplistic.

The book is divided into three Sections, let's say three big chapters. The third section is the shortest because it has to do with a few practical steps applicable to everyday life. The first section deals with Science and presents some summary ideas that are needed in the second and third sections. The second section is the largest one because it contains a lot of information, nevertheless it also remains a manual a practical handbook. It gives a taste of the Triadic System.

Many times we present or write words like Consciousness with a capital C, etc. This is to emphasize the universal meaning and not the finite or limited meaning. Figures are presented in as few number as possible. Each Section has its own References.

About Author

Menas C. Kafatos *is the Fletcher Jones Endowed Professor of Computational Physics at Chapman University and the Director of the Institute for Earth, Computing, Human and Observing (ECHO).*

Author, physicist and philosopher, he works in Quantum Mechanics (QM), General Relativity and cosmology, the environment, climate change, associated natural hazards, and extensively on the philosophical issues of consciousness, connecting science to metaphysical traditions. He leads collaborations of his research teams with a number of universities and international institutions, including the University of Athens, the California Institute of Integral Studies, Korea University, the Chungnam National University, other Korean universities, Kyung Hee University, Ewha Woman's University, and Seoul National University, NASA centers particularly the Jet Propulsion Laboratory in the US, the National Observatory of Athens (NOA) and the Institute of Geodynamics of NOA, Academia Sinica in Taiwan, Chiba University in Japan, University of Basilicata in Italy, collaborated in Earth observing satellite and climate projects of the PRC and participated in Astrophysics and remote sensing collaborations at Pune.

He holds seminars and workshops for individuals, groups and corporations on the universal principles from science and philosophy for well-being and ever-expanding human potential. At George Mason University as Chairman of the Physics Department and Director of the Institute for Computational Sciences and Informatics and subsequently at Chapman University as Founding Dean of the Schmid College of Science and Technology and Vice Chancellor and currently as the Director of the Institute for ECHO, he promoted and *promotes interdisciplinary educational and research projects, leading many multi-million $ and multi-year sciences, data sciences and remote sensing grants. His doctoral thesis advisor was the renowned M.I.T. professor Philip Morrison who studied under J. Robert Oppenheimer and participated in the Manhattan Project. Subsequently at Cornell and then at M.I.T., Morrison became involved in quantum physics, high energy astrophysics as well as the Search for Extraterrestrial Intelligence (SETI).*

Kafatos is author or editor of 25 books, including The Conscious Universe (Springer, 2000), Looking In, Seeing Out (Theosophical Publishing House, 1991), Living the Living Presence (in Greek, Melissa, 2017; and in Korean, Miruksa Press, 2016), Science, Reality and Everyday Life (in Greek, Asimakis 2019; in English 2020; and present monograph Prisma Press, Auroville, India, 2024), and is co-author with Deepak Chopra of the NY Times Bestseller You are the Universe (Harmony/Random House/ Penguin, 2017, translated into many languages and in many countries). Also, Chopra, D., Kafatos, M.C., (2023) You Are the Universe, in Korean, Kim Young Sa Publishers. Co-author of two book chapters as well as co-editor of Kafatos, M.C., Banerji, D., Struppa, D.C., Editors (2024), Quantum and Consciousness Revisited, Nalanda Consciousness Network & DK PRINTWORLD, India.

Publishes in many in high impact journals and top reputation journals such as two in each, Nature and Scientific American, as well as in Astrophysical Journal, Journal of Climate, etc. **Research Gate** *in its database lists 445+ publications under his name with RG Score: 43.17; h-index: 52. Refereed Publications: 350 which include refereed journal articles, refereed book chapters, and refereed published proceedings.* **Books: 25. Citations: 9,821 +** *(Research Gate,* up to 2021). **Highest Citation of single article: 500.** He *is recipient of the Rustum Roy Award from the Chopra Foundation, February 2011, which "honors individuals whose devotion and commitment to their passion for finding answers in their field is matched only by their commitment to humanity"; Member, Board of Trustees, Universities Space Research Association (USRA), 2006-2008; Member, OCTANe Board, 2010-2013; Member, American Hellenic Council, 2011-present; President, American*

Hellenic Council, 2014-2018; President, Friends of Sivananda Ashram Yoga in the Caribbean, 2020 - present; IEEE Orange County Chapter - Outstanding Leadership and Professional Service Award, October, 2011, etc. He has been interviewed numerous times by: U.S. national TV networks (ABC, KCBS, Voice of America), Korean and Greek TV networks (KBS1 in Korea; ERT, SKAI-Eco, in Greece; PIT in Cyprus; TV and radio stations in Crete, cretalive, tv CRETA, Krete tv), national and regional newspapers and radios in Korea, (Hankook), Greece (Kathimerini, Eleutherotypia, Ethnos, Patris), and the United States (National Herald, OC Register, L.A. Times, Washington Post, Atlanta Journal, Korea Times).

Foreign member of national academies, including the Korean Academy of Science and Technology, the Romanian Academy of Science, and collaborates with Academia Sinica.

In the past, he wrote blogs for the Huffington Post and the San Francisco Chronicle. You can learn more at http://www.menaskafatos.com

***TEDx Athens** Speaker, 28 May 2022 on "Exalted Mind, Science, Reality & Everyday Life".*

Contents

Triadic Science

In the first section, the first part, we give some basic concepts from science, mainly from quantum mechanics. We follow the sequence and material encountered in other books *Living the Living Presence* (in Greek), in presentations and lectures, with some new features and publications that came out recently.

The science we are referring to is the science of the future. It contains all of today's sciences, accepts them as ways of studying the Universe, which we can say, are studies of the visible universe, but going beyond the natural and biological sciences to what they refer to as the so-called "outer" world of objects. The quantum universe is the first part of the future science. However, there is also the world of the individual, which includes the mental levels, the intellectual levels, the biological levels. The Science of the Future will include the mental world and the biological connection to the mind, the "components" of what we believe constitute a human being, forming the second part or aspect of the Triadic Science. Finally, there is the world of experiences, the third aspect of the Triadic Science. The Science of the Future will definitely address the world of experiences, what we call *qualia*, the experiences of the individual human being. Therefore the Science of the Future will be Triadic.

The Quantum Universe

The first part of the Triadic Science is the quantum universe. Everything in the world we live in, which we often take to be just the natural universe, is based on physics. And life itself is still dependent on physics, especially modern physics as we know it today. Therefore, if we want to know how the practical aspects of our daily lives are interconnected, we need to understand, to study the laws of the universe. We can of course blindly follow the daily routine, what most of our fellow human beings, the "educated", or the "successful" ones believe, in the hope that they know what they are doing! But do they really know? How do we know what we know? Truth is not based on beliefs or systems, whether social, economic, etc. Those systems that humans create for different reasons or purposes, their knowledge is finite and constantly changing. Truth must ultimately include science, yet go beyond science. We believe that the principle is to know that the truth must be based on universal principles, on universal laws, starting from quantum mechanics.

As modern physics is based on two main branches, quantum mechanics, QM and the theory of relativity, particularly general relativity, GR, it makes little sense to rely exclusively or too much in the older physics, the so-called classical physics, formulated in the 17th, 18th and 19th centuries, beginning with the great scientist, or natural philosopher as the term was used back then, Isaac Newton, except as limiting case that gives a general sense of the observable, "out there", external cosmos around us. Or at least it's worth understanding the limits of older physics. As I emphasize again and again, quantum mechanics is not just a theory, it has a very practical nature and applications. The general, fundamental laws, on which quantum mechanics is based, are applicable to the entire physical universe; we can even claim that these laws apply everywhere, in the natural, mental, biological world, they apply to philosophies and sources of religions, in art, everywhere. Quantum mechanics is based on the same general laws that apply to our daily activities. Quantum mechanics explains how the sun shines, how photosynthesis works, it has much to do with the existence of life itself. The laws of QM are applicable to life, spirit, music (and why not?) and emotions. Thanks to these laws technology is flourishing with "smart" mobile phones, thanks to quantum mechanics we have advanced and useful technologies. It is clear that quantum phenomena, such as entanglement, non-locality, the role of measurements and the observer effect, to name a few, will find wide applications in future technologies, beyond the current advanced state of affairs. If we consider the term Nature or the Universe as the grand sum of the physical, biological, mental, social, and spiritual realities, then should it not be important to observe and understand what are the general laws applicable to Nature? These laws are obvious in quantum phenomena, which is why QM is so useful beyond its mere applications. To emphasize, the laws of nature based on QM also apply primarily to daily life, they apply everywhere! Not only in smartphones. They apply to you and me, yes to you and me!

Based on these laws of the quantum universe, it is time to realize that our way of thinking based on the classical world of the five senses is limited or even obsolete, in other words, the world of the senses is very limited. The quantum universe is about principles like interconnection, non-locality, the complementary truths that form the whole, a reality in which the observer is extremely important, the boundaries of observation being equally very important. There are no two worlds, the quantum world and the (seemingly classical) world of everyday life. There is only one world, the Universe that embraces everything and does not follow the classical laws that we thought in earlier times were universal.

Swāmī Vishnudevananda identifies important aspects of thoughts: "Thought is an object. Thought is an energy. We can transfer thoughts, give thoughts or take thoughts back. Thought has shape, colour power and energy". In some sense, these are known as qualia, the qualities of consciousness.

Now let's talk specifically: All phenomena (especially quantum phenomena) are based on a small number of principles or Laws. Despite the fact that there are other specific laws of physics (such as the laws of electromagnetic phenomena), nevertheless, according to the quantum teachings, we conclude that the minimum number of such genera Laws is three. These general Physical Laws work in everything we experience, in what we feel, in what we do, in what we encounter and how we describe the physical universe around us, the objects we observe, other human beings, other life forms and the Universe beyond the physical, the general Universe, which includes everything, includes the mental world, the world of dreams and deep sleep, which includes and in fact is based on the inner worlds:

The first is the Law of Complementarity (a more common or everyday term would be *Integrated Polarity*, while *Complementarity* would be a more scientific term). According to this Law, opposites create pairs, providing unity through differentiation, the process of creating differences not being permanent. Its logical assumptions are "Yes and/or No" rather than the dialectical expression of opposites "Yes or No". This becomes more obvious when we look at Figure 1: A quantum behaves as a particle or a wave, which are opposites, therefore we can say that while a quantum behaves *either as a particle or a wave*, according to quantum choices; its nature is a *particle and/or wave* (but never *both* under the same circumstances/ conditions). To emphasize the point, in the quantum world, quanta are either localized (in which case we name them particles), or vibrations/ oscillations (in which case we name them waves), but never both under the same conditions.

Complementarity

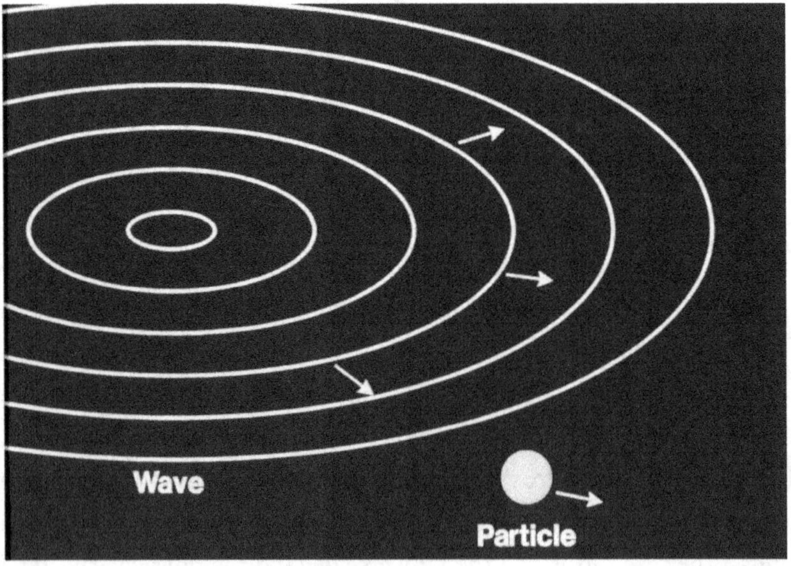

Figure 1

The second Law is that of Recursion (a more common term would be *Universality*, while *Recursion* would be a more scientific term), a Law found in many traditions such as the tradition of the hermetic philosophers, who were followers of the legendary ancient philosopher Hermes Trismegistus, coming to us from world-wide ancient and more recent traditions during the Renaissance, mystical traditions, all stating the universal principle "As here, so elsewhere". "In Heaven, as on Earth". Without the Second Law, there would be no logic, there would be no mathematics, in fact there would be no science. Because as scientists we believe that the universe can be studied, using logic and mathematics. In other words, to make sense of everything around us, from "here" we can learn about "there" (and of course vice versa).

The third Law is that of Interactivity (which implies a *Flow of interactions*; the scientific term would be *Creative Interactivity*): "everything flows" as the great ancient Greek philosopher Heraclitus used to say. In QM, we are not talking about specific objects. Quantum field theory, the modern theory that replaced the original QM, predicts that everything flows, everything vibrates or fluctuates. These interactions, these pulsations, affect everything, both those things that are "near us" and those that are "far away" from us. Every being, every person, every object in the universe interacts with something else. Through these interactions, we exist as human beings, as societies, as individual lives, we invent and develop economic, social and physical sciences, and as human beings we experience relationships, happiness and love. We notice that the number three is very prevalent in the world, it exists everywhere in nature, such as e.g. the three fundamental Laws. Such as the triadic set of subject-object and the relationships between the two, to which we will devote a lot of attention as we proceed in our discussion.

Quantum mechanics makes some perhaps equally striking suggestions: First and foremost, reality, whatever that term means, is based on the wave nature of *quanta*. Quantum mechanics does not provide clear and unique answers, it only gives us *probability of events* to happen. In other words, quantum mechanical equations do not tell us when something will happen, only the probability of what *may* happen as a function of time. Quantum mechanics is *inherently* probabilistic, with *unpredictable results* rather *than just random effects* (or events): The laws of physics probabilistically predict potential outcomes of future results, not actual results.

Because of the nature of the *wave function* itself that describes any quantum system, "particles" are entangled with other particles, everything in the universe are entangled. In addition, quantum mechanics tells us that (physical) reality is neither given nor external, and that it exists with specific attributes only when there is an observation. Quanta have no properties until observed, or as we shall see further down, information is associated with quanta, the universe is participatory. We do not just observe, we are involved in what is called "reality", as physicist John A. Wheeler used to say. Because the quantum world is essentially probabilistic, Nature also has the freedom to "answer" our questions in an unpredictable way for us. We can only calculate the probabilities of possible outcomes, that is, what *can* be done, not what *will* happen. Classical physics is essentially causal and leaves no room for the free will and energy of the individual human. The classical world does not exist, as it is a rough or only approximate approach, but it seems to be true because of our senses and our mind.

In QM, measurements are of course tied to fundamental aspects of the theory (Wigner, 1983). As such, measurements are tied to laboratory outcomes which test theoretical predictions and can be used for further experiments and refinements for better understanding of both theory and practical aspects (see for example Jordan and Siddiqi, 2024).

A fundamental aspect of the nature of quanta, providing deep insights, is the famous "double slit experiment". In dealing with quanta, whether they are tiny electrons or whether they are chunks of light energy, one can probe their nature by means of this slit experiment. Specifically, to determine whether the electrons are particles or waves, quantum scientists can set up and perform an experiment by sending a beam of electrons to a surface passing through two thin slits. A similar experiment can be set up by using light. The interesting thing about this experiment is that as the beam passes through the slits and falls onto the wall behind it, the electrons sometimes appear to exhibit a behavior characteristic of particles, sometimes exhibit a behavior of waves. This has to do with the observer's individual choices during the experiment. In order to determine what "really happened", scientists have the choice of placing a mechanism of observation next or a detector to each of the slits, in order to detect the electrons or in the case of light the photons (quanta of light) passing through. In other words, to understand what happened, did an electron pass through or didn't it pass through a particular slit? The electron acts as a particle only in this case. This occurs as if the electron "realizing" that it is being observed, decides to change its state! But how can this be explained? When in a separate case, quantum physicists do not place detectors in the two slits, then the wave pattern is observed. Starting with Niels Bohr, quantum physicists concluded that "observer mediation brings to light quantum reality". In this particular case, the electron can be either a particle or a wave, what will happen depends on the observational conditions, it depends on the *observer's participation.*

Quantum Universe

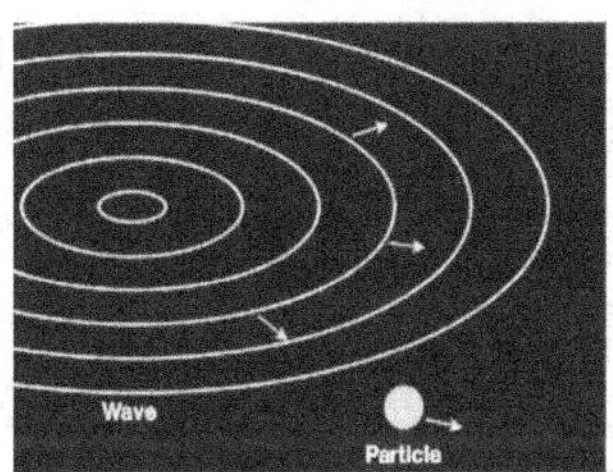

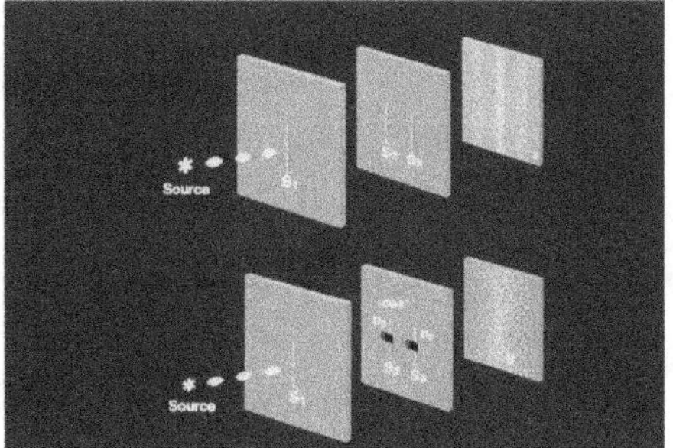

- **Wave-Particle**
- **Entanglement**
- **Non-locality**

Figure 2

Therefore, a central problem for QM today is the so-called "observer effect". It is a great mystery because quantum mechanics considers that a particle/quantum, e.g. an electron, does not even exist, it does not exist as a particle, until one observes it. Of course, even if no particular (human) observer is present, there are other observers. One may ask, what happens if we remove all the observers? The conclusion, as we shall see, is that everything is the Observer, one cannot remove the Observer. We're not talking about just humans or human observers. This would imply an anthropocentric world. We are talking about any observers, which includes human observers.

However, our present position following the study of various quantum phenomena is to extend the interpretation of Niels Bohr/ Werner Heisenberg / Wolfgang Pauli / John von Neumann, the so-called Copenhagen Orthodox view, a position we term the "Enhanced Orthodox interpretation" which actually enhances the currently popular, dominant Orthodox Copenhagen Interpretation. The persistence of the Copenhagen Interpretation, specifically pursued by Bohr, that local observers are the only active participant-observers, is not the entire picture when we try to explain the results of some quantum experiments. This step is actually indeed a very big step.

One possible weakness of our claim that there is only one Observer, as the interpretation of the so-called "quantum eraser" experiments seems to show, is that such an interpretation still requires that local observers exist to first begin the experiment, and also other observers exist, not necessarily the same, who return to see the results at the end of the experiment (the so-called "Grandfather" at the beginning and the so-called "Granddaughter" at the end, these referring to the observers). This is not an inherent weakness, nevertheless it is a weakness that requires more explanation than provided by standard QM. Note that the Grandfather sets up the experiment, at any time *before* the experiment starts and therefore his actions have no impact on the results. Passive Observation does not in any way influence the end results of the experiment. Only what happens in the Active Observation box of Figure 3 influences the results of the experiment.

Such a view opens up the overall metaphysical picture which the present book follows, which would be difficult for the "western" types adhering to realism to accept. We see this phenomenon, denying that there is basically a vast world beyond the senses, is occurring again and again.

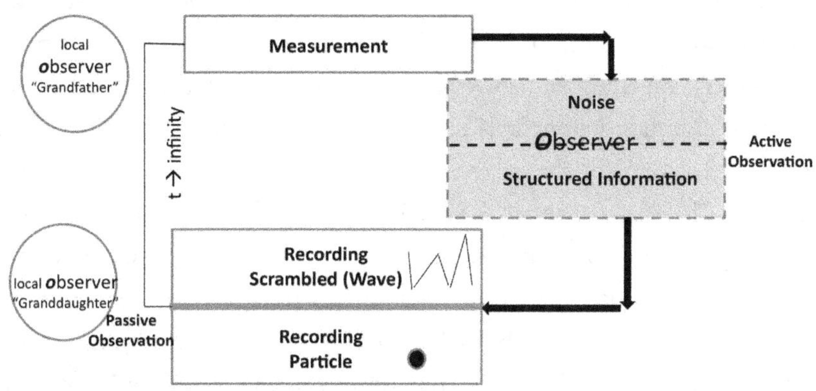

Trinity: Observation, Measurement and Recording
Figure 3

This basic view of quantum mechanics for the central role of observation is in agreement with Orthodox mysticism, with Buddhism, with the stochastic traditions of *Vedānta*, which speak of the consciousness of the observer. In quantum mechanics the observer is not a specific human observer. The observation itself is important. Thus, in modern quantum mechanics, we say that there is no real self (in other words the ego) in the sense of a material substance which does not change. We all are *processes* (following the Third Law). The great Heraclitus taught that you cannot enter the same waters twice. This is what quantum mechanics says in a different way. Observations transform objects into reality and the Universe changes! So it is really strange that quantum mechanics which involves quanta/ particles is essentially a theory of the mind. Today we can express the statement that quantum mechanics is a theory of how the mind generates what we call reality!

In quantum mechanics, emptiness and fullness are complementary truths. The quantum void is "empty", but it is also completely "full", complete, as far as it is possible, because it contains all possible outcomes. If however, we believe that only the void exists, then we are wrong.

Therefore, since the universe is participatory, observations always have some kind of a limit. Boundaries are created through the process of observation, the limitations that ultimately manifest through our minds. But here is the question: How can the quantum world, which is non-local and entangled, appear classical, Newtonian, full of discrete objects? We need to identify the answer as to the role of the mind. Therefore, ultimately, human existence is very important. The covering of quantum reality by the mind conceals the underlying consciousness and blurs experience. Curiously, quantum mechanics is basically about the nature of the field of the mind. Psychologists, cognitive scientists, psychics should all study quantum mechanics. In the same way, quantum physicists should all study psychology and cognitive sciences!

The mind

The second member of Triadic Science is the human being, whose main characteristic is the mind. Because what we refer to as a human being has so many, multiple, aspects and levels many of which we will deal in the second part of the book, let us concentrate here on some questions concerning the mind. And a few questions about the human body.

The Science of the future will be down to earth, very practical, and will not reject Philosophy in the general sense of the Unity of the Universe and the role of the mind. Perhaps the only divide or separation that exists between humans and the world is the creation of the mind? If this is the case, then the mind is at the same time a great tool for studying what we are, but also a great trap. The human mind may trap the human being herself/himself, we are trapped by the mind, by thoughts, which are our own creations, then we are trapped in them and then we call what we think and feel as the "truth".

Inasmuch as describing some basics of quantum mechanics, now let's explore its deep implications for the nature of mind. The mind with its mysteries and control is a central aspect of the universe as Swāmī Sivananda (1994) eloquently discusses. In any case, what is ultimately the mind? Modern neuroscience, if it even accepts the mind, it claims that it is the result of some physical processes of the brain's neurons, that is to say, it is the result of biological evolution.

This way of thinking holds the view that after an unknown amount of time had passed, the mind gradually developed, beyond the simple complexity of the brain circuits, and reached the point of modern human, the point where the human wonders about the origin of the mind itself. But how exactly this happened or is still happening, has never been discussed or answered, except with general arguments based on metaphysical beliefs, even if they are called "scientific". However, if we look at the physical brain, there is no physical place, there is no area for the mind to be found. In the best of cases, we find that when certain experiences occur, then certain areas of the brain are light up, what are known as the natural correlations of experiences.

What would happen though, as it is indicated in many systems, if the human mind were part of the universal Mind, part of the cosmic Consciousness? We can let the current scientists who attribute everything to material existence, disagree about the nature of the mind, since we know, you and I know, that we have a mind and we know that our mind is not limited by any physical part of the human body, even the brain (except of course in extreme circumstances) and can travel to the most distant parts of the universe, or remain in deep reflection, right here and now.

It is a distinct possibility and we have plenty of indications that the mind cannot as such be limited in location in any particular part of the physical body, including the physical brain, limited to a set of neurons. Some modern scientists claim that the mind is non-local and not confined to the physical body. This recent view makes sense in the quantum nature of the universe. As the so-called quanta can be interconnected non-locally, crossing unimaginably large, even infinite distances of space and time, then why should the mind, which is much more sophisticated, not extend everywhere, possibly even beyond space and time? We of course are not saying that the brain or the neurons play no role in the mind. They are giving us human-based understanding, experiences, are tied to qualia and much more. But this does not mean that the mind, which interfaces with "physical" entities is itself limited in physical space. Future Science will examine the practical applications of the deep truth associated with the human species, what we call the mind. Moreover, if we want to know more about the nature of the mind, we need to know some things about time, about experiences in general because mind and time seem to be closely linked.

If the mind is a construct of the biological body from which it was "evolved" somehow, evolved through the physical brain, let us examine what the "physical" body itself is. Well, the physical body is a structure composed of some one hundred trillion cells, the majority of which are not even human cells! The non-human cells are bacteria, foreign cells, and are numerically superior by at least 10 to 1 in relation to human cells. One could therefore argue that the ("physical") body is a colony of non-human, one could call tehm alien, cells. As I often say, our body is a colony of foreign cells, they are inside us, in our bodies, in our organs, in our brains, everywhere. Our body is in some sense an alien planet of one hundred trillion cells, bacteria, viruses, seeds, etc. But of course, all these cells work in harmony with one another and with the human cells, paradoxically giving us the sense of being a human being. Instead of (these foreign) cells mutually annihilating each other and annihilating the entire colony we call our "body", they work seamlessly in the "body". When very few of these cells behave abnormally, for example through mutations, they create havoc, In reality, we are a cell colony, a symbiotic colony of many trillions of cells.

And then the question, how do all these cells create the mind? This is the big question.

Experiences

The third member of the Triadic Science is what we call experiences. Experiences begin with the basic question, who am I? Am I just a body destined to die sometime, consisting of 100 trillion cells? Or am I something more? I'm definitely made up of the human body and the mind. But is *this* all? What is it that which always remain the same? When one looks oneself in the mirror every morning, one does not see a hundred trillion cells working together, even though the physical body is basically that. What we see is a person, with the same eyes, hair, like yesterday and the day before yesterday, all these parts of our body and the invisible mind, elements that constitute our familiar "I". What does not change, even though all our cells change over in 7-9 years (actually most of them even sooner, some over days), is the so-called awareness of our existence. Let us focus on what is the basic nature of Awareness. First and foremost, it is existence or entity of being and before this existence, there was more existence. In fact, it is not possible to remove existence from existence! The second is the ability to be conscious (to be aware of oneself and of other objects). The third aspect is completeness, fullness per se. Awareness is truly complete, nothing else is needed for awareness. *It just is.* This is the basis of all existence. This fullness can be called bliss.

How can one "conquer" the Trinity of Awareness, Consciousness? There is no need to conquer it, no need to approach it, as it is always there! All basically spiritual traditions converge on the same message, to basic the question: "Who am I?" The primordial feeling we feel when we look at the stars at night or when we hear the sound of the ocean waves or see the beauty of Nature, seems to be lost as we grow and become busy with everyday life. Nevertheless, there is one thing that is certain in life, no matter what life is, as life evolves and gives rise to countless forms, it is that life eventually "ends". Is this the human destiny that we cannot escape? Deep inside we know instinctively that this cannot be so. But our society tells us otherwise, we are busy with life and lose the sense of wonder, the sense of wonder that springs out of existence itself. This is how we end up as "an adult man", "an adult woman", repeating the same patterns again and again, often mechanically, until one day we die.

In order to gain a sense of our self, let us hear what Socrates, the great philosopher, had to say: "One thing I know is that I don't know anything". We think we know, but is this true? What do we know? When we don't even have the answer to the question "Who am I?" Socrates taught that one has to start from the acceptance of one's own finite mind, and then, as he taught through his method of asking questions and through the process of discourse-dialectics-with his students, a new truth would emerge. In this way, knowledge is a dynamic process, expanding and constantly changing. The other great Greek philosopher, Heraclitus, taught that "everything is progressing, nothing remains stationary. This teaching is identified with the teachings of Buddha, the way of the Tao, the teachings of the great teachers of India, as we find them in non-dualism: In the (monistic) *Vedānta* and in *Kashmir Śaivism,* and in complete agreement with quantum mechanics: this process, this flow, is what is *true*, what is *real*.

We have different levels of existence and different senses of time. There is wisdom at every level. There is the wisdom of the Earth. There is the wisdom of the Sun. There is the wisdom of the human body. There is the wisdom of animals and plants, the wisdom of the living Earth. Everything works together, often in the form of competition but also in cooperation, and *all* those together is exactly what we call the universe.

Experiences cannot be identified or localized in space time, as objects are. The qualities of experience are the qualia. The experience of red color, a thought, the experience of time, all are qualia. Qualia is a Latin term meaning "what kind". Qualia give us impressions that there exist "external" objects but the experiences are internal. Experiences like color, smell, time, space etc. make us realize that qualia are the fundamental building materials of the universe, not the particles, atoms or physical objects as all these are also qualia. The idea is basically in line with quantum mechanics. It is the universal manifestation of the third Law of Nature (Creative Interactivity). Objects, space-time, everuthing, they are all experiences. The Science of the Future will study the essence of experiences.

A key experience we have is the experience of time: It is perhaps the basic "problem" of human existence, as we are all apparently bound by time. It seems that we are born as human beings, we live our lives and at some point we leave this world. At least we believe that there is the flow of time, beginning, middle and end. But where is time? Modern quantum mechanics claims that the linear time of classical physics actually does not exist. Time is more cyclical, repetitive. Cycles over cycles of time. The cycle, the circle, has neither a beginning nor an end. In ancient Greek thinking there were two types of time, one called *Time* (in Greek *Chronos*) and it was about external phenomena. The second one was called *Kairos*, it was subjective time, recording our inner experiences. One can then consider that *Chronos* and *Kairos* are complementary truths of a generalized notion of Time (which refers to changes) i.e. they are application of the First Law (Complementarity). The two types of Time apply to all levels, i.e. they are application of the Second Law. Moreover, they are involved in innumerable interfaces, i.e. they are application of the Third Law, either through observing/recording of "the generally accepted time lapse", characterizing events, or through observing/recording of the internal qualia/ experience of time, the subjective experiences.

The ancient Greeks, the Hellenes, named the beginning and the end of Time as alpha and omega. For Orthodox Christians, Jesus is the alpha and the omega, the beginning and the end, the cosmic Being. Buddha understood the meaning of time. The time that exists is in the final analysis the Present, which is beyond time! In the second and third sections of this book we will see more things about time and the experience of time in our lives. The conclusion is that only the eternal Now exists.

Philosophical systems of the ancient Hellenes covered a great variety of issues related to humanity, the role of humans in the cosmos as well as ethical and right living, as discussed by Plato, Aristotle and the Stoics. The ancient philosophers as exemplified by the life of Socrates, were not necessarily interested in cosmic questions, rather what is the right living, the proper and just life for humans. Philosophical systems of the Hellenes remain beacons for thousands of years and will remain for eternity. As such, science is not a technical endeavor, it connects to the deepest aspects of humans. For the Science of the Future, we remember what the great Plato declared: "Science separated from Virtue, is evil."

References

[1]   Aristotle *Nicomachean Ethics*.

[2]   A. Aspect, "Bell's Inequality Test: More Ideal than Ever", *Nature*, **398**, p. 189 (1999).

[3]   J. Bell, *Physics* **1 (3)**, 195 (1964).

[4]   D. Bohm, *Wholeness and the Implicate Order*. Routledge, London (1980).

[5]   N. Bohr, *Atomic theory and the description of nature*. Cambridge University Press, Cambridge UK (1934).

[6]   D. Chopra, M. C. Kafatos, *You Are the Universe*, Random House (2017). And in Greek,

[7]   D. Chopra, M. C. Kafatos, Μ.ηνάς Κ. Καφάτος, *Είσαι το Σύμπαν*, Εκδόσεις Π. Ασημάκης, Μετάφραση Ρ. Καρακατσάνη (2018).

[8]   D. J. Chalmers, *The Conscious Mind: In Search of a Fundamental Theory*, Oxford University Press, Oxford (1996).

[9]   A.S. Eddington, *The Philosophy of Physical Science*, Cambridge, Cambridge University Press (1939).

[10]   A. Einstein, B. Podolsky, N. Rosen. *Phys. Rev.* **47**, 777 (1935).

[11]   D. Harrison, *Complementarity and the Copenhagen Interpretation of Quantum Mechanics*. UPSCALE. Dept. of Physics, U. of Toronto (2002).

[12]   W. Heisenberg, *Physics and Philosophy: The Revolution in Modern Science*. George Allen & Unwin, London (1958).

[13]   W. James, *The Principles of Psychology*. New York (1890).

[14]   Μ.ηνάς Κ. Καφάτος, *Βιώνοντας τη Ζωντανή Παρουσία*, Εκδόσεις Μέλισσα, Μετάφραση Σπύρος Πετρουνάκος (2018).

[15]   A.J. Jordan, I.A. Siddiqi, *Quantum Measurement: Theory & Practice*. Cambridge University Press (2024).

[16] Kafatos, M.C., Banerji, D., Struppa, D.C., Editors, *Quantum and Consciousness Revisited*, Nalanda Consciousness Network & DK PRINTWORLD, India (2024).

[17] M. Kafatos, Th. Kafatou, *Looking in, Seeing out: Consciousness and Cosmos*. Quest Books, Wheaton, IL (1991).

[18] M. Kafatos, R. Nadeau, *The Conscious Universe: Parts and Wholes in Physical Reality*. NY: Springer-Verlag (2000).

[19] M. C. Kafatos, "Fundamental Mathematics of Consciousness", *Cosmos and History: The Journal of Natural and Social Philosophy 11(2)*, 175-188, (2015). http://www.cosmosandhistory.org/index.php/journal

[20] A. Narasimhan, M.C. Kafatos, "Wave Particle Duality, the Observer and Retrocausality", *Quantum Retrocausation III*, Daniel P. Sheehan (edit). AIP Conference Proceedings, 1841:040004-1, 9 (2016).

[21] Plato, *Timaeus & Kritias*, several published versions exist.

[22] E. Schrödinger, *What is life?:The physical aspects of the living cell*. Cambridge: Cambridge Univ. Press (2001).

[23] Swāmī Sivananda, *MIND Its Mysteries and Control. The Divine Life Society (1994)*.

[24] H. P. Stapp, *Mindful Universe: Quantum Mechanics and the Participating Observer*. The Frontiers Collection, 2nd ed. Berlin, Heidelberg: Springer-Verlag (2007).

[25] H. P. Stapp, *Mind, Matter, and Quantum Mechanics*. Springer-Verlag, Berlin, Heidelberg (2009).

[26] N. D. Theise, M. C. Kafatos, "Fundamental Awareness: A Framework for Integrating Science, Philosophy and Metaphysics", *Communicative & Integrative Biology* **9(3)**: e1155010, DOI: 10.1080/19420889.2016.115501000-00 (2016).

[27] J. von Neumann, *Mathematical Foundations of Quantum Mechanics*, translated by Robert T. Beyer. Princeton University Press, Princeton, NJ (1955).

[28] J.A. Wheeler, in *Some Strangeness in the Proportion*, ed. H. Woolf, Reading, Addison-Wesley Publishing Co. (1981).

[29] A.N. Whitehead, *Process and Reality*. New York: The Free Press (1978).

[30] E. Wigner, "The Problem of Measurement", In: *Quantum Theory and Measurement*, J.A. Wheeler, & W.H. Zurek (Eds.), Princeton University Press, Princeton (1983).

Triadic Reality

Introduction

The philosophies of idealism in all times, all traditions, have the same basic message of existence: The Universe, humans, everything, are not physical objects. Mind, consciousness, are not based on physical processes (although they appear to be). What we will be referring to alternatively as Consciousness (with capital C), or as Awareness (capital A), is beyond time and space. Consciousness, Awareness refer to universality while consciousness, awareness (with small c and small a) refer to "individual" situations. In the philosophy of idealism, after all, the individual is universal. There is only Awareness which is Existence, without a "second" state.

Idealism does not mean disconnection from reality. Exactly the opposite. We will see that imagining a world that exists *without* consciousness, does not make sense. If it did, how would we even know? As we have seen, the most consistent explanation of the most successful scientific theory, quantum mechanics, states that the mind, consciousness (individual or global, although ultimately they are both the same) is a key "player" in the processes underlying everything experienced in the world. "The Universe is participatory," declares quantum and cosmological physicist John Archibald Wheeler. This is also the meaning of "You are the Universe", the book by Deepak Chopra and Menas Kafatos.

Often in our use of words, when we refer to consciousness, we refer to knowing something as an object. While by the term awareness, we usually refer to as an internal process. But after all, as we shall see, the two are *exactly the same.*

There are three universal powers of consciousness: Will, Knowledge and Action. In order to act one requires knowledge and will, and knowledge presupposes will (to learn). Everything in the universe is created, sustained and eventually disappears. Even the universe itself. These are three universal functions of the universe, and everything in the universe. We find them in thoughts, in the functions of life, in cells, in planets and stars, everywhere. There are two more functions, one is to hide the true nature of things. The last one, revelation (or recognition), is to reveal the reality of one's own true nature of existence.

What is it that which always remains the same? When one looks in the mirror every morning, one sees a person, the same as the day before, the same eyes, hair, all these elements that constitute the familiar "I". It is the so-called awareness of our (own) existence. This is the message of all spiritual paths. This awareness has three aspects, the "ego" (subject), the "other" (the object, which could be the "ego" in self-awareness) and the relationship between the two, the most important of all, which could be formulated in the following sentence: "I Am That", constituting a trinity, the triad of existence. Me (I) and the Universe, and relationships between the two.

Let us focus on the basic nature of Awareness. First and foremost, it is existence, entity. "Before" existence, there was more existence. It is not possible to remove existence from existence. The second is the ability to be conscious (of oneself, of other objects). The third aspect is completeness per se. These three aspects constitute the first *Trinity*.

Awareness is really complete, it does not need anything else. This is the basis of all existence. This fullness can be called bliss. Basically all spiritual traditions converge on the same message: liberation, or convergence with one's own essential being or existence, begins with asking the following question: *Who am I*, truly, really?

Kashmir Śaivism or the Triadic System

Ancient Śaivism known as the Triadic System (*Trika*) of philosophy and its more recent and most developed form, as developed in Kashmir from the

8th to the 12th century, is a set of philosophical teachings applicable to daily life, its purpose being liberation (*mokṣa*) of the individual human being, with profound practical consequences for humanity. Śaivism flourished in Kashmir, in a beautiful nature setting with high mountains and cool waters, developed and built on the basis of the tradition of the ancient *Vedas*, and *Vedānta* the scriptures which followed the Vedas. However, Śaivism seems to be much more ancient.

The Śaivism we are interested in (because there are other forms of Śaivism, such as partly non-dual and partly dual as well as purely dualistic Śaivism) is perhaps unique to all non-dual systems because it accepts the reality of all that exists. It does not deny the existence of the Universe, but instead considers that the Universe is as real as the infinite Śiva. This Śiva is Supreme Awareness. Thus, Śaivism does not go against science, on the contrary, of all non-dual systems, it comes closer to science and especially to quantum mechanics with the paramount role of observation. From now on when we say Śaivism, we mean the Triadic (*Trika*) System of Kashmir, which was developed in a short time, but in a very active, golden age, in Kashmir and reached its zenith in the 12th century. Masters-Sages/ Philosophers, who most often were Teachers as well, such as: Vāmana (779-813 AD), Vasugupta (875 - 925), Utpaladeva (student of Vasugupta, 900 - 950), Kallaṭa (also student of Vasugupta), Bhāskara, Somānanda, the great Master-Sage/Philosopher, Teacher, Abhinavagupta (950-1020) and Kṣemarāja (the most important student of Abhinavagupta), have left us a rich tradition that aims at the highest goal of human evolution, who we are, how to live daily in the spiritual awareness of the greatness of Reality. The Teachers-Masters of Śaivism lived in the daily society of their time, many of them had families, others were monks, yet they approached daily life, from the highest level of perfection. Śaivism does not mean disconnecting from reality, quite the contrary, it is intended to show that we can live everyday life by developing the inner vision, the wisdom aimed at the highest Reality.

Of course one can say that it all sounds great but what do we do every day when our fears, survival worries, life challenges drown us? What do we do when our own mind does not let us rest during challenging or even the worst conditions, which don't seem to be rare in the present times, to find rest, peace of mind? This the point, Śaivism is not just a theory, it is *also* a prescription of right action, through higher understanding.

As we mentioned, Śaivism does not deny religion, does not replace, it does not replace religion. Moreover, it does not replace science. I believe that Śaivism helps us to discern the deeper meaning of religion, if we are religious believers; to discern the deeper meaning of science, if we follow objective research and more importantly, it may help us to see that these two "poles" are not opposite to each other but that they are *complementary*. Of course, it also does not go against any philosophy, ancient or modern, for it says that after all, despite all the *apparent* dualities, all systems of philosophy and different understandings, all arise from the same Source, which is universal Conscousness, Awareness. With Śaivism, we may perhaps understand, for example, what the great philosophers, Socrates and Plato were teaching and what Aristotle taught, based on them, to achieve the right measure in one's life.

But now let's get to the key point, the key question: *Who am I?* Śaivism answers unequivocally, with complete clarity: We are *Awareness*, pure Universal *Consciousness*. All experiences are inherent components of transcendental Reality, which is universal, non-dual Awareness. The individual, however, is subject to a *misunderstanding* of the nature of Reality, living in dualism, in seemingly different situations, *separated* from her/his true nature. The human mind, though of the same nature as pure Consciousness, has "forgotten" its Non-Dual nature.

All observers, all humans with their different experiences, all individuals, are but the *One* Consciousness. This seems to be in line with the indication of the existence of a Universal Observer that we have seen some quantum experiments indicate in the first part of the book when we examined the issues of quantum observation, measurement and recording of data of propagating quanta in the laboratory.

According to Śaivism: Supreme Śiva (*Paramaśiva*, or *Supreme Consciousness*, the Absolute Reality without even a trace of difference, the *Union of All*). Plato would call it *Essence*, supreme *Substance*. Supreme

Śiva consists always/is the *Light*, Being. It consists of the Supreme Being (*Prakāśa*) and the Consciousness of Being (*Vimarśa*). Prakāśa is the Eternal Light of Existence, without which nothing can appear or nothing can done at all. It's Śiva. *Vimarśa* is the Śakti, the mirror through which Śiva realizes his infinite powers, his magnificence, his beauty. Śiva and Śakti, they always go together, they are not separate.

Absolute Freedom (*Svātantrya*) is the main attribute of Śiva, Absolute Freedom and Absolute Power to do whatever there is and to create everything. It has infinite powers. However, three are primary, Will, Knowledge, Action (*Icchā, Jñāna, Kriyā*). Śiva is the Perfect I-ness, the Supreme *Brahman* (Cosmic Principle), the Supreme *Paramātmā* (Supreme Self).

According to the other great non-dual system of Vedānta (i.e. the scriptures whose sources are the ancient Vedas): Brahman, the Absolute Being is composed of the Trinity: Being (*Sat*) - Consciousness (*Cit*) - (Divine) Bliss or Fullness (*Ānanda)*. In the non-dual Vedānta developed by the great Teacher, Master-Sage Śaṅkarācārya, Brahman is only Prakāśa or jñāna (or Light, direct Knowledge), but without Vimarśa (power of action).

Triadic (*Trika*) system contains many triads or trinities, the most important of which are:1) Śiva, the absolute, undifferentiated Being 2) Śakti (Universal Energy), also known as *Citi,* Śakti, *Parāśamvit,* Ānanda, etc. 3) Individual, bound soul, *Jīva* or *Nara.* Another trinity consists of 3 states 1) *Parā* - the highest, supreme 2) *Parāparā* - the middle state consisting of identity within differences, and the last 3) *Aparā* - which is the state of difference, i.e. we and Śiva, "we are separated". Another triad consists of the three fundamental Powers: 1) Will, 2) Knowledge, and 3) Action. Another triad consists of the three bodies (the fourth one is, as we shall see, the basis of the three bodies we experience as ordinary human beings). The three levels of sound (the fourth one is also the basis of the other three levels of sound), etc.

In the triads of Śaivism, we notice other triads from science, namely the three Universal Laws. The triadic relation between the subject, the object and the relations between them (which is of course the basic relation in the

Triadic System) and the most basic relation in the concept of qualia, etc. The Trinity, of course, also exists in religions, it is the Three-headed God in the Orthodox and Catholic Church, parallel to Christianity, in Buddhism, in Vedānta, and all great spiritual paths. In the Orthodox tradition, the Holy Trinity consists of the three Persons that are One: Father, Son, and Holy Spirit. In the Hindu pantheon, Brahma (the creator), Vishnu (the one who preserves) and Śiva (the one who dissolves).

To emphasize what we already discussed, as these are paramount concepts: There are also three universal powers of Awareness: Will, Knowledge and Action. To act requires knowledge and will, and knowledge presupposes will (to learn). Trinity is the basis of all things, the triadic aspects which in the end are of course the same as the Unit, non-duality is the same as the Trinity. In the non-dual Vedānta, the relation of Reality to the Triune Nature consists in that Reality consists of Existence, Awareness and Completeness (spiritual bliss), i.e. the Three are One.

Plato in *Timaeus* states (the text from *Timaeus* here is in italics, whereas text where we provide comments or explanations, is in parentheses): *God did not create the soul after the body of the world, that is to say,* as we (humans) *tend to believe, because he would never let the senior be governed by the younger, when he would connect the soul and the body. We just tend to be saying things in a random way that largely characterizes our existence. God placed the soul higher on the scale of birth and virtue* (see below, the levels of reality, *tattvas*), *to hold dominion and govern the body. Its composition was as follows: From the indivisible and always unchanging Essence or Substance* (we would say in analogy from the level of Paramaśiva, the Supreme Śiva) *and from the divisible and changing in the physical bodies Essence or Substance* (Śakti, which in its Supreme form as Parāśakti, Parāśamvit, is the same as Paramaśiva. Śakti in the changing world of tattvas, as we shall see, appears as *Prakṛti*, Nature), *composed a third kind of Essence, intermediate consisting of both.* (i.e. here we note: all tattvas, from the level *māyā*, to the level of Earth element. In *Timaeus* we see the analogy with the three states of the Triadic system, *Parā* - the highest, *Parāparā* - the middle state consisting of identity within differences, and the *Aparā* - which is the state of difference). *In the case of Identity and Difference again, following the same principle, composed intermediate mixtures, consisting of the indivisible (non-dual) and the divisible (dual) parts of the bodies. Taking then these three components he developed them into one form, forcing the Difference, which by its nature is hard to mix* (and ending in the three "defilements" which cause division and limitation), *to unite with the Identity and then the mixture of the two* (i.e. Śiva and Śakti) *to be united with the Supreme Essence* (i.e. Paramaśiva which is identical with the Parāśakti).

The Triadic System incorporated Tantric and Buddhist influences but eventually became an autonomous, majestic system which nonetheless is non-dual, as Advaita Vedānta is, but differs in key steps. After all, both systems aim towards the same purpose, Liberation. Like these other systems, the main purpose of Śaivism is liberation (*Mokṣa*).

Here are the main aspects of Śiva: Absolute Existence ("passive" aspect of reality), while *Śakti:* Absolute Awareness ("dynamic" aspect). However, there is really *no difference* between *Śiva* and Śakti, they just *appear* different. From our point of view, we can look at *two* sets of "polarities" realizing they form the *One Reality.* These two different sets of polarities in Trika Śaivism are as follows: Śiva: Universal Consciousness (*Cit*); without any form, absolute Emptiness, the underlying Reality; Beyond the mind and the senses; represented as "Masculine" (realizing of course that such terms apply to ordinary existence, not to the Absolute); *transcendental Peace,* without any movement, eternally the same, without a beginning or an end; the Great Lord of the Universe, the Transcendental Self (what ancient *yogīs* referred to as *That*); Absolute Subject, the One without a second. Personified as the Divine Teacher, the Divine *Yogī*, Śiva is the Lord of all spiritual paths. Śakti: Consort of Śiva, Absolute bliss, (Ānanda). Creates all the infinite forms out of Her own Divinity; Completeness, the Power of Creation, Maintenance and Re-absorption, the Origin of all universes; "acts" as the mind and the senses although She is beyond everything; represented as "Feminine", the dynamic Energy, the source of infinite energies, changing all the time; "inside" the Universe, rather She is the Universe or most likely an infinite set of universes, levels of experience and countless worlds. The

Great Power, personified as the *Mother* of all, lives in all creatures, in all beings and supports them all.

Very briefly, Śiva's "family" includes his consort Parvati's (who is of course yet another name for Śakti), includes one of their two sons, the deity-man-elephant *Ganesh*, one of the most beloved forms of the ancient Pantheon of India. Ganesh is as beloved today as he was in the past because he removes the obstacles any student faces in her/his spiritual path. Lord of knowledge and the mind. Ganesh has a real existence in millions of people around the world. Whatever one says about Ganesh, it will be too little.

Kashmir Śaivism *Trika* literature is divided into: 1) *Āgama śāstra:* Revealed scriptural writings of Śiva. 2) *Spanda śāstra:* Doctrines on Śakti as universal *Vibration.* 3) *Pratyabhijñā śāstra:* Philosophical works on *Recognition.* Some great works include: 1) *Āgama: Śiva Sūtras, Vijñānabhairava* 2) Spanda*: Spanda Kārikā,* and 3) Pratyabhijñā: *Pratyabhijñā-hṛdayam, īśvara Pratyabhijñā Kārikā, Paramārthasāra, Tantrāloka*

Liberation in the philosophy/practice of Śaivism: Consists in residing in universal Awareness, the inner Light. It is the Transcendence beyond body-mind states. As we saw, Trika Śaivism is both a practical and philosophical system, holding the view of *many in One,* the non-dual system transcending all differentiations. Observation and Stillness in the *Śiva* state. The *Now,* non-dual state, which is beyond space and time, beyond all circumstances which change, the state of *Being.*

Fetters

To know liberation, we must first know what the "bonds", what the "fetters" are. The basic fetters in the Triadic System are the three defilements, or the three impurities called *Malas.* The term does not imply uncleanness, it means the basic reasons that a human being does not feel her true Nature, her true Self. The defilements or impurities refer to the *ego-self (the obscure aspects of illusion).* It is a term that refers to forms of self-concealment and they are three: 1) The first one is the feeling that we are incomplete, different, separate from Consciousness. It is the ignorance of Reality, of who we truly are. It is a primary and deep sense of separation, the subtlest feeling, hard to pinpoint but always there. This impurity is the real, big "enemy". Utpaladevā in *Īśvarapratyabhijñā Kārikā* describes the first defilement, *Āṇava Mala*, as the one that covers freedom, separates universal Consciousness from *Āṇu*, the limited creature. It limits Will, *Icchā* Śakti, cut off from the global stream of Consciousness. It is the subtlest impurity, the particular internal impurity of the individual. The second defilement consists of unclean knowledge, the sense of duality. It is subtle but less so than the first one. *Māyīya Mala* is a limited condition caused by *māyā*, illusion. It is the awareness of differences caused by the limitation of Knowledge, *Jñāna* Śakti. The last one, *Kārma Mala*, is the impression of pleasure and pain. It arises from past actions (lives?) that produce pleasure or pain. It is the least subtle of the three defilements. Kārma mala is a limited state caused by the subtle impressions of previous acts, the *vāsanas*.

The three impurities create pockets, "veils", or *shells* which are covering the soul. These shells are: *The Body, the Energy body, the Mind, the Intellect* (also known as *the subtle mind) and the Bliss of deep sleep.*

Non-dual Vedānta and the *Sāṅkhya* philosophical system regard ignorance as the cause of bondage. While Śaivism holds the view that liberation is the awareness that the universe and our self are indivisible and complete, *Being is Everything that exists.*

Śaivism, according to the great contemporary 20th-century Śaivite Master Swāmī Lakshmanjoo, explains that ignorance is of two kinds: Ignorance in which one does not know his/her true nature in *samādhi*. It is removed by the grace of teachers who themselves live the experience of the scriptures, and by meditation. Spiritual ignorance in which one does not know the philosophical truths of Śaivism. This is removed by studying the monistic texts Śaiva that explain the reality of the Self. It is based on thought and is developed through (pure) intellect.

Śiva Sūtras

The *Śiva Sūtras* scriptures were according to tradition revealed to Master-Sage-Teacher Vasugupta, in a specific location at a well-known mountain in Kashmir. In one version, the revelation was given in a dream. In another version, Lord Śiva pointed to Vasugupta where he should go to find the sūtras on a rock at the base of Mahadeva Mountain in Kashmir. He went as directed, found the rock, he touched it and the rock turned over revealing the Śiva Sūtras. In a third version, a *Siddha* (perfected being) revealed the scriptures. In the present work, we often use the accepted and widely-used term *sūtra* to denote concise statements or statements of spiritual rules or truths to be studied, contemplated upon and apply to life. The sublime Śiva Sūtras hold the foundation of Śaivism. They consist of three sections of different *upāyas* ("means" to do *yoga,* to achieve Liberation): I. *Śāmbhavopāya Yoga* ("path of *Śiva*", Continuous spontaneous state of Will of Awareness of *Śāmbho,* the great Śiva; the Self is realized through the will of the Master). They are 23. II. *Śāktopāya Yoga* ("path of *Śakti*", the path of knowledge). They are 10. III. *Āṇavopāya Yoga* ("path of *Āṇu*", *the* individual human being, using a variety of means such as contemplation, breathing, sensations). They are 46.

It is to be noted here that the three paths are available depending on the state of the student, the seeker. They should not be taken as higher or lower, better or worst, they all ultimately lead to the state of *Śiva.*

Sūtra I.1: caitanyam ātmā or caitanyamātmā

Consciousness (is the) *Self.* In Sanskrit the "is" is implied, therefore this statement can be expressed as: "Consciousness-Self". The actual Sanskrit term for Consciousness refers to Luminous Awareness, the Light of Consciousness. Consciousness means more than just (conscious) awareness, it has the absolute freedom of will, knowledge and action. The first sūtra gives the highest teaching, the highest understanding. By itself is a complete path, a complete yoga.

Sūtra I.2: jñānam bandhaḥ

Limited knowledge is (the cause of) *bondage*. Ignorance of one's real nature is bondage. Ignorance gives rise to āṇava mala. All rooted in the impurity of *māyā* or illusion, it creates bondage, the mark of which is obstruction of true knowledge.

Now Swāmī Lakshmanjoo presents two different but complementary ways to interpret the first two sūtras, reading them separately, or combining them. In the second way,

Sūtra I.1 and Sūtra I.2: caitanyamātma+ajñānam bandhaḥ

The first two *sūtras* when reading them separately, the second sūtra focuses on "knowledge is bondage", limited knowledge is the cause of bondage; while when the first and the second sūtras are *combined*, the second sūtra now states that *no-knowledge is bondage*. What do these two statements really mean? Are they contradictory? No, they are not. In the first case, (attempts) to *know* the *limited* universe (and confusing it with ultimate Reality) constitute bondage. In the second case, *not knowing* the real unlimited Self (Reality), proclaimed in the first sūtra, *is bondage*. Both statements are correct from two perspectives.

The four *Mahavakyas* (the term meaning "the great statements", or "the great truths") of the Upaniṣads convey the same message as caitanyam ātmā. As Jaidev Singh and Swāmī Laksmanjoo explain, the main characteristic of the universal being (Śiva) is absolute Freedom. It is not only infinite, it has no limited or has unlimited ability to act, etc. (like other "things", e.g. the physical universe). The complete freedom of the divine Being is the fundamental nature, the ultimate cause of the universe. Which means (are you ready for this?) that there is no explanation for the universe, there is no "reason" for us to find out, no "why". This belief has profound implications for science and ultimately results in the Socratic "one thing I know is that I don't know anything".

The first Sūtra of śāmbhavopāya, *caitanyam* ātmā, offers the highest truth. Swāmī Lakshmanjoo and Jaidev Singh provide detailed explanations on this point: Reality is (unlimited) Conscience/ Consciousness. The independent state of higher Consciousness is the reality of everything. And finally, absolute freedom is the nature of Consciousness. We see the first Sūtra referring to Śiva himself.

As soon as we start judging and using names, or worse when criticizing others, we have strayed from the *basic meaning of* Śaivism, then we enter the world of duality. At this point the world is then full of duality (or more correctly "appears"), we live in duality and we call this world of existence of dualities, the "reality"

We can of course have our own views and believe that this is the truth, but actually this results from the situation we are in, which the second sutra states, *jñānam bandhaḥ*, which is, limited knowledge is the cause of bondage. Swāmī Lakshmanjoo explains both of these first two sūtras really well: It is Śiva who takes upon himself these limitations, becomes finite for the sake of divine play, while in fact remaining completely free. And liberation is nothing but remembering this truth and acting in our lives from this point of elevated understanding, beyond the narrowness of the mind.

Sūtra I.4: *jñānādhiṣṭhānaṁ mātṛkā*

The ground of knowledge is *Mātṛk*ā. She is the ground of both superior and inferior knowledge. Mātṛkā is the un-understood Mother or Power of sound inherent in the alphabet, which is the basis of limited knowledge, in the form of Āṇava, Māyīya, and Kārma Malas. But mātṛkā is also the basis of expanded spiritual knowledge. Written texts provide and record knowledge, however in ordinary awareness, all knowledge is limited and the cause of bondage. The fetters, the Malas, are the instruments of bondage. Knowledge of the Self is infinite Conscious Awareness, beyond the fetters.

Sūtra I.18: *lokānadaḥ samādhisukham*

The bliss of the world is the bliss of *samādhi*. The delight of the *yogī* in abiding as the knower in *both* object and subject (in a sense the "outer" and the "inner") is the bliss of *samādhi*. Ultimately, the world itself is full of bliss, when one focuses inwardly to taste the nectar of Consciousness and its Divine Energy. The words "inner" and "outer" are mere conventions, the world is One, it is the manifestation of Śiva. As such, the *yogī* regards the entire world as one with one's own Self. This delight is experienced inwardly and in a sense spills over to everything that exists, this blissful awareness is communicated to everyone around, to others.

The means of Energy ("path of *Śakti*"), using deeper meditative practices, *Śāktopāya*, constitute the second book. This is the path wherein usage of *mantraḥ* is the main approach. *mantraḥ*, the Divine Word, leads the seeker to inner Conscious Awareness. In a sense *mantraḥ* is the word of Parāśakti, the Divine Creative Energy.

Sūtra II.1: cittam mantraḥ

By intensive awareness of one's identity with the Highest Reality enshrined in the mantraḥ and thus becoming identical with that Reality, the mind (*cittam*) itself becomes *mantraḥ*. Citta ponders the highest Reality. One then becomes identical with the mantraḥ, Consciousness expresses Itself in the divine mantraḥ, the mantraḥ leads to divine Knowledge.

Sūtra II.3: vidyāśarīra sattā mantrarahasyam

The luminous being of the perfect I-Consciousness inherent in the multitude of words whose essence consists in the knowledge of the highest non-dualism, the highest usage of language, constitutes the secret of mantraḥ the highest understanding. All mantras consist of letters and *mātṛkā* (the power of sound inherent in the letters of the alphabet) is the very form of Śiva.

Sūtra II.5: vidyāsamutthāne svābhāvike khecarī śivāvasthā

As spontaneous supreme knowledge emerges, the state of movement in the infinite unlimited expanse of Consciousness, the infinite Sky of Consciousnes which is the *Śiva* state, is realized, we can say this one expanse is the state of *Khecarī*. More precisely, the state of *Khecarī* is the state of supreme embodiment of *Śiva*. The individual then becomes one with *Śiva*, supreme Consciousness.

Sūtra II.6: Gururupāyaḥ

The Teacher (Guru) is the means. This teacher (who is a Master, has attained Self-realization, which itself is knowledge of Reality) is the only one who can help the spiritual aspirant, the spiritual student. Such a Teacher lives in the constant awareness of being one with Śiva. The Guru empowers the *mantraḥ*, the Guru is the grace-empowering function of the Lord. With the inner opened eye, the disciple then realizes that the Guru is not necessarily a bodily existence, the Guru is here and everywhere, is one with Śiva. The Guru is Consciousness (Śiva) and the divine Śakti.

The individual path through a variety of practices of yoga, *Āṇavopāya*, constitutes the third book. The individual through different yogic practices advances in spirituality and brings the restless mind to the abode of the Inner Self. In a sense, the third book is the path of the individual, the individual through yoga practices, advances in the spiritual path and is led to Śiva. *Āṇavopāya* is a longer path, however even the long path leads to Śiva.

Sūtra III.1: ātmā cittam

The individual Self is mind. From a limited point of view, *ātmā* identifies with the intellect, the mind and the ego. As such, the individual self identifies with the mind. The individual being is entangled in the wheel of repeated birth and death.

Sūtra III.4: śarīre saṁhāraḥ kalānām

Dissolution of the various parts of the *tattvas* in the body should be practiced by *bhāvanā* (spiritual "burning", which is the successive removal of negativities). There are five circles, from gross to subtle, one entering the other, which constitutes the dissolution of all constituent parts.

Sūtra III.7: mohajayād anantābhogāt sahajavidyājayaḥ

Achieving an all pervasive conquest of delusion, delusion referred to by the term *māyā*, specifically by destroying the impressions of delusion, one masters the natural, inherent knowledge of Reality. This conquest is boundless, it has no limits. The seeker then enters or becomes the Divine State.

Sūtra III.9: nartaka ātmā

The *dancer* in the field of universal dance (which like a dance changes while the dancer is retaining harmony) is his *Self*, the Being of universal supreme Consciousness. Such a one who has realized his/her essential spiritual nature is none other than the universal Self, the Lord. This Self is only an actor on the world stage. In an allegory, the Self is *the* dancer.

Sūtra III.12: dhīvaśāt sattva-siddhiḥ

Through the higher spiritual intelligence, the realization of the Light arises (with all associated powers) of the Self. The *yogī* experiences that she/he is acting.

Sūtra III.20: triṣu caturthaṁ tailavadāsecyam

The fourth state is the state of meditation. The other three states as we will see below, are also called "bodies" and they are: Ordinary waking state; state of dreams; state of dreamless sleep, "deep sleep". This fourth abode permeates the other three states like oil.

Sūtra III.25: śiva-tulyo jāyate

Such a *yogī becomes like Śiva* (by going beyond the fourth *turya* state, the state of meditation without thoughts, to the state beyond, *turyātīta*). *Turya*: All of the energies reside there but are not in manifestation, they are without distinction. A *yogī* remains unagitated by all these energies which in the past constituted bondage. It is the state of complete thoughtlessness (Swāmī Lakshmanjoo). *Turyātīta*: It is that state which is the absolute fullness of the Self, filled with consciousness and bliss. It is the last state. Beyond here there is nowhere to go. It is the "unlimited and unexplainable supreme totality" (cf. Swāmī Lakshmanjoo).

Sūtra III.42: bhūta-kañcuki tadā vimukto bhūyaḥ patisamaḥ paraḥ

Then on the actual ending of desire, he uses the body of gross elements merely as covering. As such, for him the five elements are mere coverings. At that very moment, he is absolutely liberated. Such a being becomes liberated, a state where the so-called individual is pre-eminently like *Śiva*, the perfect Reality.

Tattvas

In terms of the etymology of the word *Tattvas*, the word is composed of two syllables, meaning "this" from *tat* and the *tvam* suffix: Different meanings regarding the term Tattva is found, such as: category, truth, substance, reality, principles. They are the fundamental elements in every Indian philosophical system of schools. Or we may say that it refers to the term qualia, the essence of experience (the singular and the plural in Sanskrit is the same; there is no "s".

Tattvas of the universal experience (5)

The first 5 Tattvas are the levels of Perfection, of Non-Dual experiences. They are shown here in Figure 4 below, and although the (Cosmic) Subject (the *I*, Ego with capital E, *not* the limited ego) looks separate from the (Cosmic) Object (*That*), the different levels show different relationships between the two poles, within Non-Duality. At the first two levels, *Śiva-Śakti* are united. At the third level, the Subject (*I*) has the greatest emphasis. At the fourth, Object (*That*) has the greatest emphasis. At the fifth level, the two are in perfect balance, the *I - That* and the reverse, *That – I*, have the same equivalence. In these relationships, the verbs "am" for the Subject *I*, and "is" for the Object *That*, are implied grammatically but not written (as in the first sūtra, *Consciousness* (is the) *Self*, the verb *is*, is implied).

Reality of Self & Illusion of ego

5 "Cloaks"

<u>Perfection Illusion</u>

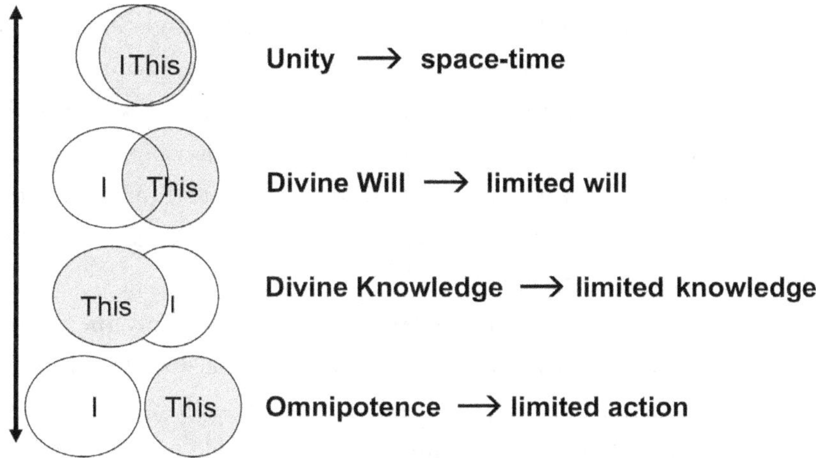

Figure 4

Awareness has three aspects, ego (individual subject), that (individual object, which could itself be the "ego", for example in the case of self inquiry) and the relationship of the two, this being the most important of all, which could be formulated as the following sentence: «I am That». These three aspects constitute a triad. The triad of everyday world becomes the Trinity at the 5 perfect levels: In other words, the relationship between the Absolute Subject, Being (Śiva) and the Absolute Object (Śakti), does not cause separation between them. The first two Tattvas are One with some slight variation: If the two sides of a coin were to be both on one and the same side, we would have some analogy. The two sides of the "coin", the relationship of first two tattvas is I/That. Above the union of I/That, or Śiva/Śakti, there is the *Absolute, Paramaśiva* which we have seen is identical with *Parāśakti*. This level, does not exist as separate, it is not included in the Tattvas. We cannot even try to "describe" it by analogy. Perhaps the closest we can say is that it's everything and nothing!

At the third level, after the first two, at the Tattva of the universal *Will*, the Cosmic Subject and the Cosmic Object continue to not be separated, but as we saw the Subject holds more emphasis than the Object (and this kind of relationship is expressed as **I** – That, the bold indicating the dominant entity). At the fourth level, at the Tattva of *Omniscience*, they are still not separated, but here the Object is more emphasized (this kind of relationship is expressed as **That** – I, the bold again indicating the dominant entity). At the fifth level, at the Tattva of *Omnipotence*, they are still not separated, but here the two are in perfect balance (these kinds of relationship are expressed as **I** - **That**, or **That** – **I**, where now both the Subject and the Object have equal weight). The 5 levels are, as we saw, the levels of Perfection, and the number 5 is the number of Śiva.

In Figure 4, the left column shows the perfect levels of Unity, the 5 perfect levels of Śiva. Next to each, the limited levels of Illusion are also shown (the arrows → indicate that the perfect levels become, through illusion, limited). At the Illusion levels, the relationships **I** – **that** still hold but now the subject and the object "appear" as individual, small, bound entities.

Figure 5

The figure above gives additional information for the 5 pure levels (tattvas) to that given in the simpler version of Figure 4: At each level, the following information is provided: The name of the corresponding Divinity; the name of Higher Beings that exist there; the mantra of the level; and the name of the Divine Power. For example, at the level of Divine Will (*I – That*) the names are: *Sadāśiva*; *Mantra Maheśvara*; *aham idam*; and *Icchā Śakti.*

Following, "below" from the fifth Tattva, the restrictions of limited existence begin: The tattva of the limited individual experience (6) which is the Illusion (māyā) and her 5 veils, her followers, the *bonds* that bind humans to finite existence (Figure 6). They are called *kañcukas.* They are the *tools to bind individual existence.* They are in order: *Māyā*, the Illusion covered by 5 veils, the 5 shackles of individual existence. However, for Śiva, they are his 5 ornaments, or the bracelets of Śiva. They are: 1) *kalā*: Limited ability 2) *vidyā*: Limited knowledge 3) *rāga*: Desire (for limited things, driven by the senses) 4) *niyati*: (finite) Space 5) *kāla*: (finite) Time.

Without the primordial illusion of māyā and its 5 veils, Śiva would not be the Universe. Śakti, the Great Mother, possesses the (mysterious or unknown) creativity, the ability to make the Infinite finite. This is the great mystery, the mystery that the Absolute, although it never ceases to be Absolute, resembles a finite, individual existence in space, time and with limited forces.

māyā & kañcukas

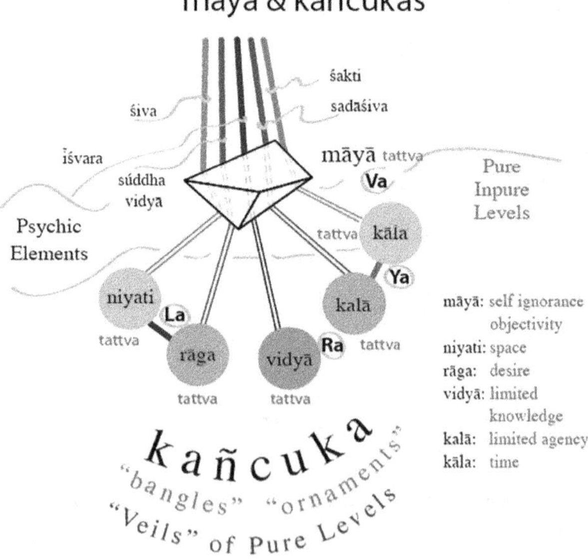

Figure 6

Following these tattvas of limitations (the veils of Illusion), 6 in total, the following tattvas are found in sequence:

#Tattvas of a limited existence, the *individual being*, on the one hand, and *Nature*, on the other hand (2). The individual in this case is not just a human being but any individual being (including animals, etc.)

#Tattvas of mental function, the *intellect*, the (objective) *ego* and the *mind* (3).

#Tattvas of sensory experiences (a total 15, as follows): The abilities of perception of the senses, the internal capacities of the sensory organs (e.g. eyes and the ability to see) (5). The abilities of action, kinetic organs (5). The primary, subtle, elements of perception (5). And, finally,

#Tattvas of matter, the "gross" or coarse elements, specifically: Ether, Air, Fire, Water, and Earth (5).

The 25 Tattvas of Śaivism below the tattvas/levels of Illusion (*māyā*) and her 5 veils, the *kañcukas* (Figure 6) are shown in Figure 7. These 25 tattvas being below māyā and controlled by Her are "impure": The individual and Nature are now separated. Māyā and the kañcukas are, on the other hand, "pure-impure" as the individual (subjectivity) and Nature (objectivity) are not yet "formed" and as such not, not yet separated. However the tools for separation and binding are there (the impure part). But when we realize that Māyā is after all nothing else than the Great Śakti, the Divine Mother who creates, maintains and re-absorbs all existence, everything there is, then we recognize the Divine Play of Śiva and Śakti (the pure part).

Traditionally in the Indian systems wherein the Tattvas describe the different levels of reality, there are 36 Tattvas. However, Abhinavagupta added two additional states *Mahamāyā and Guṇa tattva*. The three qualities of Nature that are aspects of *guṇa* are *sāttva, rājas and tāmas*: *sāttva* (meaning goodness or light) is purity. The second quality, *rājas,* is the dynamic aspect, often associated with the limited mind. The third quality *tāmas* denotes dullness, inertia, ignorance, associated with the ego. These qualities in the distant past were initially in balance, in perfection, however as evolution took hold of all beings, in a way because of impressions and what led to actions or past Kārmas, one or more of them ended up dominating, particularly the mind and the ego.

The qualities of Nature are depicted in Figure 7, the 25 limited tattvas. These qualities apply everywhere, including most important for us. We should not expect to understand them at first glance. Slowly the experience levels that the tattvas represent will "speak" to us, as we study them more and more, we will understand them over time because they are within us. Why would we need to understand them? Because they hide the secrets of human existence, life and death, the secrets of the Universe as the human being is but a microcosm of the Universe.

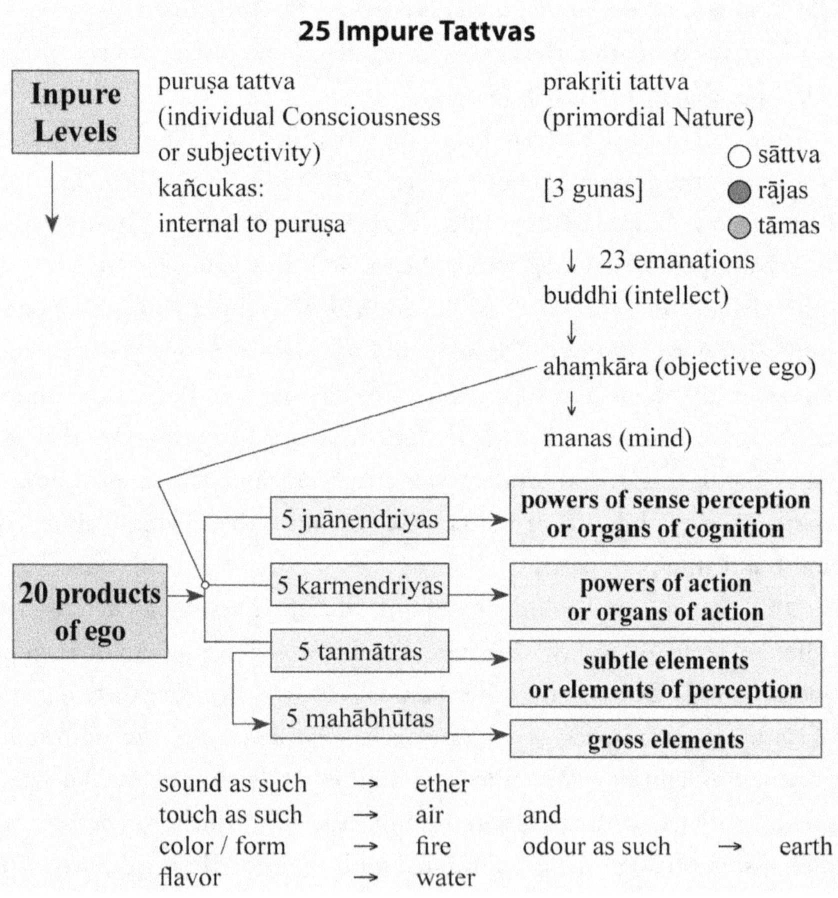

Figure 7

There is other information in the 36 tattvas of the Triadic Śaivism System, such as the connection of the fields of existence (36 tattvas) with the letters of the Sanskrit alphabet (50 in total) and the numerical order that the letters appear at the various levels (e.g. the first letter *a* is 1, the second letter ā is 2, etc.). The connection is made to the cakras, where the sounds of the letters, the tattvas, and the arithmetic order are in relation. But above all the 36 tattvas, above all the cakras, beyond all experience, is the Absolute, the ultimate, greatest Mystery, Paramaśiva = Parāśakti. Although *That* is beyond all tattvas, yet it is in all tattvas, giving with the Light of Existence and Consciousness, life and existence to infinite beings in infinite worlds. *Thou Art That.*

Liberation and Cosmic Dance

After encountering the "bonds", the "fetters", māyā and her 5 veils, the kañcukas, which "appear" to shackle one into individual existence, let's now look a bit closer at *Liberation*. After all, Śaivism is all about *Recognition* of our true nature, who we truly are, and *this* Recognition is Liberation, the recognition that we are pure Consciousness. But then one may ask what is "liberation" from "what"? Since we are already Śiva and Śakti, what is the problem? The problem is that yes, we are *That*, but *we don't feel* that way. That is, we have forgotten it. What matters is what keeps us from feeling our Reality, what we already are. These are the shackles. As I say in my workshops, "know your enemy." The enemy is the *Malas* that actually do not exist, as they are the subtle and gross impressions of what is a false separation, the Āṇava, Māyīya, and Kārma Malas. They don't exist in the same way that the (small) ego (me as individual) ultimately does not *really* exist. Yet they are very real, because they control us. They are like non-existent ghosts that we cannot escape, and we consider as the world "out there" as separate from us. When we recognize our true nature, they disappear.

Śiva «plays» the universal game, the universal play on the cosmic theater, He is the great *Actor*. He is dancing the universal dance, in the form of *Śiva Nataraj*. Where are we in all this cosmic dance? One of the most *basic secrets* of Liberation is to see, to realize the five Actions of the cosmic dance: The cosmic dance consists of the five Actions that are: Creation. Support. Dissolution (or Reabsorption). Hiding (of what is our true nature). Finally, Revelation/which is Release (revealing what is hidden, *That* which is already there). We perform these five actions all the time, all our lives! Everything in the Universe performs these five actions, or processes! From Śiva, to the smallest particle, the smallest cell, the smallest grain of sand.

Other interesting and important concepts are: The number 5 is the number of Śiva. The number 3 (e.g. the trident of Śiva) is the number of Śakti. Three plus five is eight, the *eightfold* way!

Bodies

A human being has three "bodies", that is, three states of his finite, conditional existence: The first, is the physical body that corresponds to the awakened state of ordinary "everyday" life. It is the world governed by classical sciences, physics, biology, it is the shared world of experiences that we have. The second body is the ethereal, the body of dreams or the dream state. The third body is the causal body, the state of deep sleep without dreams. As countless yogis teach and practice, Swāmī Vishnudevananda also connects Grace to physical existence, "if there is God's grace you will be able to do intense sadhana to purify your physical, astral and causal bodies". Without purification, sadhana is controlled by the limited mind.

The fourth state, the fourth body is the immortal Presence. It is always there, it supports the other 3 bodies, we approach this fourth state in deep meditation, we go "within" and then we "come out". Known as *Turya*: All actions reside in Turya but do not occur, they are indistinguishable. The 4th state remains unaffected by all these actions. It is a state of complete silence, without any thought.

Then there is the state of supreme intuition *beyond, Turyātīta*: It is not a "state" in the ordinary sense, it is the state of absolute fullness of Śiva/Self, full of Consciousness and Bliss, it is the abode of Existence, Awareness and Bliss. There is nowhere else to go. It is the "unlimited and inexplicable supreme Sum".

The three bodies, or the three states of the (relative) of being, are shown in the following diagram. Through meditation, we dive through the deeper levels of the Absolute, to the Transcendental pure Existence and beyond (Turya →Turyātīta). This is where "memories" of limited existence melt into *That*.

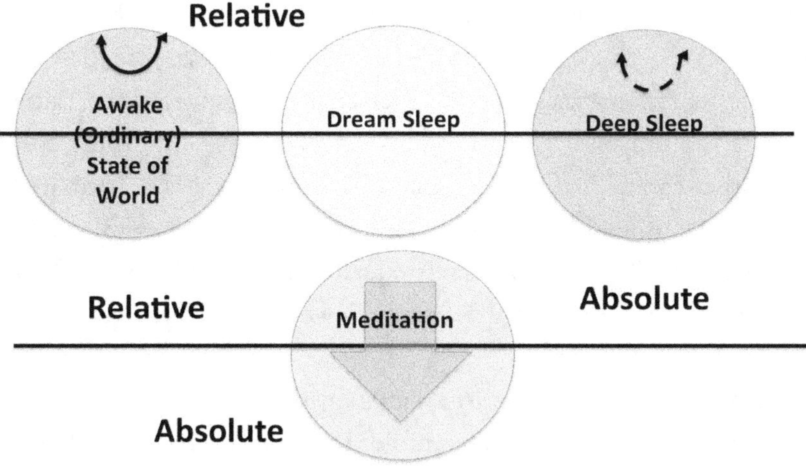

Figure 8

Pratyabhijñā-hṛdayam

The scripture *Pratyabhijñā-hṛdayam* means *Self-Recognition*: Its author was the well-known Śaivite Master Kṣemarāja, a student of Abhinavagupta. Recognition does not refer to the mundane matters but to the recognition of our true Reality. It consists of twenty sublime philosophical sūtras. We believe that it is linked to science through the three laws of Reality.

We get into the details of the first five sūtras here:

Sūtra 1: citiḥ svatantrā viśva siddhi hetuḥ

The interpretation of the terms is: *citiḥ* consciousness - *svatantrā* highly independent, completely free - *viśva* universe - *siddhi* achievement, power - *hetuḥ* cause. In other words, with different but equivalent interpretations, the first sūtra tells us that: Universal Consciousness through Its infinite freedom brings about the achievement of the Universe (emphasizing not only physical reality, but everything). Or, otherwise: The universe is the means of achieving the realization of free, universal Consciousness. Or, yet a third interpretation: Highly independent (free) universal Consciousness is the cause of the manifestation, preservation and reshaping of the universe.

How does the universe appear? What is the source of the universe? The answer is, the creative power of Consciousness. The first sūtra gives the underlying cause of the ever-changing universe, the cause of which is the universal Consciousness or Awareness. The three cosmic actions that create, maintain, and absorb all that exist are attributed to *Citi* (Consciousness) and nothing else. (Note that in the following we refer to Citi as "She" but of course equally valid is "It", we use the two interchangeable--*She* being the infinite Power of the Lord, *He*).

Sūtra 2: svecchayā svabhittau viśvam unmīlayati

By the power of her will, She unfolds the Universe in a part of herself.

Sūtra 3: tan-nānā anurūpa-grāhya-grāhaka-bhedāt

This (i.e. Consciousness) is made varied by the division of objects and objects, mutually adapted.

Sūtra 4: citi-saṁkocātmā cetano'pi saṁkucita-viśvamayaḥ

Even the individual (i.e. the individual personality), whose nature is (universal) Consciousness in a reduced state, nevertheless integrates the (whole) Universe in a reduced form. In some sense, this is the meaning of "You Are the Universe".

Sūtra 5: citir-eva-cetana-padād-avarūḍhā cetya-saṁkocinī cittam

Consciousness itself, descending from its (higher) extended state, becomes the (individual) mind, which is limited by the objects of perception.

Here we show in more detail additional thoughts or points about the first five sūtras: The 1ˢᵗ sūtra tells us that universal Consciousness with its infinite freedom is the (only) cause of the universe. Contrary to the way scientists view the universe as created and guided by the laws of Nature, the Pratyabhijñā-hṛdayam states that the cause of everything *is* universal Consciousness itself. But of course the ultimate principle of the (physical) universe, modern quantum field theory *with* general relativity (which yet have to be unified but most physicists believe that this unification will eventually occur), is the quantum "vacuum", i.e. the origin of everything is in some sense "nothing". At some point, what is ultimately the purpose of unified field theory, does not go against the first sūtra, without of course being able to prove it, because Consciousness *cannot* be proven (as an object) as since it is the basis of all of (*apparently* different) objects.

Here we see that the Triadic System agrees with the most basic belief of all religions, the existence of the Existence or Oneness, which the great Parmenides also believed. The quantum vacuum contains everything as a possibility. Kṣemarāja in Pratyabhijñā-hṛdayam states that the cause of all is Consciousness itself, so the "vacuum" of Being contains everything (as possibilities). But how? The 2ⁿᵈ sūtra states that by the power of herWill alone, Consciousness unfolds the universe (onto) a part of herself. That is, not all objects are "outer" creations, Citi unfolds the universe on her mirror, on her screen (that is, on herself, as She is the foundation of the universe, the Mother of the universe). But how, one wonders, does Universal Consciousness unfold the universe? Here Kṣemarāja describes that the so-called universe is nothing but the *projection* of Consciousness upon Consciousness. What appear as different things (objects, subjects) in the objectified world are only *projected* differences in the global mirror of Consciousness. The mirror, or the screen of Consciousness, unfolds the universe. Our nature is both transcendent and in the universe. The 2ⁿᵈ sūtra is an expression of Creative Interactivity. Transcendental Citi reflects what appears as the universe on her own Self, as it happens in a mirror.

But in a regular mirror there are external objects reflected on it. In the (we can say magical) mirror of Consciousness, on the universal screen (of Consciousness), there are no "external" objects although they look different because of the *apparent* division between subjects and objects, which are mutually adapted. Here the 3rd Sūtra gives us the fundamental relation between object and subject, the most fundamental relation of qualia, that between an object and a subject, on which all other qualia are based, the experiences of the "other", the "object". Without objects, there would be no subjects, without subjects there would be no objects. We have reciprocity. The reciprocity of influence between subject and object is the basis of experience. This mutual process is everywhere. It is an expression of recursion. The "other" (the object) is really the "I" (the subject), though it seems separate from me (through a mutual relationship between the *two* of us).

The 4th Sūtra gives us the high message, that the individual person encompasses the universe, integrates the whole universe, in a (seemingly again) reduced form, which the book *You Are the Universe* means. The levels of manifestation, all tattvas, exist in us, in our bodies, in our minds. Here the 4th Sūtra shows us the representation of the universal laws of Complementarity and at the same time of Recursion.

The 5th Sūtra says that Universal Consciousness itself, as it descends from its expanded state, becomes the (individual) mind, which is constrained by the apparent objects of perception. The (individual) mind, formed and being part of the vast ocean of Consciousness, is no different from Consciousness. But through "external" objects, it becomes limited. Thus this Sūtra is directly linked to the Śiva Sūtras rules we have seen, such as I.2 and III.1. The 5th Sūtra is linked to the Univeral Recursion law.

In what follows, we simply provide the translation of all other Pratyabhijñā-hṛdayam Sūtras:

6th Sūtra: The person whose nature is the (individual) mind, is subject to māyā.

7th Sūtra: And therefore the individual is one, has two forms, consists of three (malas), has four (bodies), and is of the nature of the seven groups of five (tattvas).

8th Sūtra: The positions of all philosophical systems (that is, all philosophies) are stages of the universal Self, of Consciousness (as are the Tattvas).

9th Sūtra: *That* (Consciousness), which is full of Consciousness, due to the reduction of its (universal) powers, ends up becoming a transmigratory soul (a soul born again and again) and covered by the malas.

10th Sūtra: However, even in that condition, the transmigratory soul performs the five actions as does Śiva.

11th Sūtra: These five actions take place in the form of illuminating (being conscious of) the objects, enjoying them, knowing them, sowing the seed (of limiting memories) and dissolving those limitations.

12th Sūtra: The state of transmigratory (reincarnating) soul is illusion, caused by its own forces, when it is not fully aware of the situation it finds itself (in other words, when it is not aware that itself is "writing" its five actions).

13th Sūtra: When one is fully aware that she/he is the author of the five actions, then one's mind becomes (universal) Consciousness, ascending to the state of complete expansion through inner meditation.

14th Sūtra: The "fire" of Consciousness (i.e. Kuṇḍalinī, as we will see below), although hidden when it descended to the lower levels, (begins to) partially burn the fuel of the (object) that is recognized.

15th Sūtra: By attending (great) Power, the individual makes the (entire) universe his/her own.

16th Sūtra: The state of liberation while the individual is still living, is the uninterrupted experience of oneness with Consciousness, even if the individual experiences his body and other aspects (i.e. the mind, feelings, etc.), a condition that follows the achievement of the bliss of Consciousness.

17th Sūtra: The bliss of Consciousness is achieved by expansion of the "center" (i.e. the rise of Kuṇḍalinī through the central channel, see below).

18th Sūtra: Specifically, the means of expanding the center are to dissolve the thoughts, to shrink and extend the power of the individual being, to stop the flow of incoming and outgoing vital energy (as we shall see associated with breathing), to be aware of the points at the beginning and end (the points between breathing and exhaling) and so on.

19th Sūtra: The permanent achievement of the state of *samādhi* is achieved by conviction or by contemplating one's own identity with Consciousness again and again, in the state that follows meditation, which is filled with the imprints of samādhi.

20th Sūtra: Then, by entering into the perfect I - ness (i.e. the state of Awareness, of the Absolute Self), whose nature is the power of the great *Mantra* and the essence of the bliss of the light of Consciousness, sovereignty is attained over its wheel of the deities of Consciousness, who perform every manifestation (of existence) and absorption. This is (the state) of Śiva.

With 20 Sūtras, the great philosophical Śaivite text Pratyabhijñā-hṛdayam shows us the highest Truth (as do the Śiva Sūtras) while it also gives us the practical steps of Recognizing that Truth from the highest levels of *Citi*.

Tantras

Śaivism is directly related to *Tantra*. The rich tradition of tantra likely originated in the seventh or eighth centuries, its goal is to bring the seeker in touch with the Divine Energy within, this is what the great Śaiva Master Abhinavagupta practiced and taught. Tantra involves rituals, ancient scriptures, religious practices, etc. One of its most important components is also called *Kaula*. The word tantra consists of two terms which mean tan = "do in detail" + tra "protect". The fourth system associated with the Trika System is the Kaula / Tantric System. Abhinavagupta in particular was initiated into Kaula, his voluminous writings contain many basic Tantra / Kaula truths. The "sacrifice" of worship in the Kaula, the so-called *yajna* (fire ceremony), is mainly defined as an inner act.

Worshiping Śakti as the Divine Mother, through the practices of mantrās and sacred shapes (yantras). The goal is to achieve union with Śiva, through the awakening of the inner spiritual energy, Kuṇḍalinī, by uniting Śiva with Śakti. Tantras involve Divine, revealed scriptures, with emphasis on: Knowledge, meditation, devotion to the teacher. It is the knowledge in the Tattvas and the associated mantras that protect those who resort to them. Tantras are practical texts on worship that involve: media of worship, meditation and mantra repetition. The texts of the Tantras were revealed to humans and are not connected to the ancient *Vedas*. The purpose of Tantras is both liberation and enjoyment. They show world correlations between humans, the universe, higher powers, as well as rituals.

These scriptures were lost in *kali yuga*, today's cycle of time (which is characterized by conflict but also contains means to reach the Highest), but were eventually rediscovered again by the long genealogy of the ancient fathers who passed on to their sons, from one branch of tradition, and to daughters, from another branch of tradition, thus preserving the ancient teachings. Tradition says that as a result some of the disciples were created by the Mind, as pure beings. Eventually the ancestors of Somānandanātha moved to Kashmir. Somānandanātha passed the teachings on to his students creating what is known as Kashmir Śaivism and since then the teachings have been transferred from teachers to students. Thus, Tantras and Kashmir Śaivism are very closely linked, especially to the great Master-Sage Abhinavagupta, bringing forth the Triadic System in his writings, and establishing a fourth component Kaula, with its own ancient writings.

The following passages are from the unique and sublime works, [6] and [10], author's comments are added at the end of each section on the yoga and the mind.

Yoga

Swāmī Muktānanda in his autobiography *The Play of Consciousness* [5] describes amazing inner experiences after receiving Divine Initiation, *Śaktipāta*, from his Guru Bhagawan Nityananda, ultimately culminating in

Liberation, in the merging of the individual existence to the Self of All, *Śiva*.

Yoga refers to Union (with the Divine). In Yoga it is important, it is often necessary to have a living Master. True Masters are one with *Śiva*. If one does not have guidance either internal or external, the mind can create structures and beliefs that may not correspond to Reality.

Swāmī Muktānanda, in his treatise *Light on the Path* [6] discusses Yoga, spirituality, insights, innumerable ways to proceed on the path in order to realize the Union with the Self.

Quotes from different sections in [6] identified by headings are shown in what follows:

Introduction

Self-realization is the supreme purpose and real meaning of life. When a person suffers from a sense of imperfection or lack of meaning in the material achievements of life, he longs for higher ideals and attainments.

The Grace of the Guru

What is the ultimate goal of life and how can it be attained? ... The seers of profound wisdom have said that to be entirely rid of pain, suffering, and sorrow and to attain the full measure of absolute bliss is the real goal of all beings.

Not only human being but all creatures on this earth, from the tiniest to the largest, seek happiness and pleasure.

Jñānamarga: The Path of Knowledge

Knowledge is one of the means of attaining God-realization.

That which imparts consciousness and possesses an independent power of creation as well as knowledge of all kinds—its state is known as *caitanya*.

Yogamarga: The Path of Concentration

Another means of attaining God-realization is *yoga*. Restraining the thought-waves that are continuously formed and modified in the mind is known as yoga.

As soon as the mind becomes calm and steady, an abundance of joy arises in the heart of the aspirant. This is the same as attaining the blissfulness described in Vedānta.

All intelligent people, even if they are householders, can practice yoga because it does not deny them a modest and rational worldly life. In fact, yoga is complementary to such a life and, in time, becomes a friend to it.

Bhaktimarga: The Path of Devotion

Intense love of God is known as *bhakti*, or devotion. Like *jñāna* and *yoga*, bhakti is a means to God, perfect in itself. The way of love is the sweetest, like nectar.

Siddhamarga: The Path of the Perfect Ones

Śaktipat Dīkshā

In every sect or religion there is a tradition of *dīkshā*, or initiation. The real meaning of dīkshā is "to give"- to give an awakening whereby the initiated one can have a superconscious vision of the Lord and ultimately experience his identity with the supreme Self.

The grace of the Guru is itself a process of initiation known as *śaktipat dīkshā*.

There are those who doubt the efficacy of śaktipat.

In every human being there dwells a divine energy, the *Kuṇḍalinī Śakti*.

Experiences of the Awakened Kuṇḍalinī

Some disciples experience great joy, while others become either apparently dull and stupefied, or restless.

The *Śakti* alone knows Herself, and She alone is fully aware of Her own true nature,...

Final Attainment

All the Vedas say that this Self is Brahman (*ayamātmā brahma*). If, therefore, the entire universe is nothing but the Lord Himself, it is the absolute truth that one can experience divine bliss, or Universal Consciousness, by following the path shown by the Guru.

The Nature of God

God the Supreme Being, the One without a second, is known by more than one name and there are many paths and forms of *sādhanā* to reach Him. God is infinite and so are his names.

There are also various arguments for and against the form and the formless nature of God. But the nature of the incomprehensible Supreme Lord does not need proof by arguments.

Sacchidānanda

The term *sat* indicates that which exists in equal proportion in all beings, at all times, and in all places. The *ātman* pervades everything in full measure and is the support of all.

The second aspect of God is *chit* which means cosmic intelligence or knowledge.

The third aspect of God is *ānanda,* that is, joy or bliss…says *Brahman* is bliss.

Guru and Disciple

The many different powers we see in this world are nothing but different manifestations of God.

One who guides aspirants on the divine path is called a Guru.

A True Disciple

Discrimination (*viveka*) and nonattachment (*vairāgya*) are the two main qualifications of one who wishes to be a disciple.

A person's suffering starts from the moment he is conceived.

How does a disciple attain oneness with the Guru? He becomes one with the Guru by subduing the ego and merging himself completely into the Guru through every thought, word, and deed.

Japa Yoga

Repetition of the name of God…is a secret, very profound and difficult to explain…by practice of *japa* alone a person can achieve success in his spiritual endeavor and attain perfection.

Mantra: Its Meaning and Potency

Continuous remembrance of mantra is japa.

One cannot complete the sadhana of japa yoga, that is, one cannot attain the highest fruit of practicing *japa*, if one lacks proper understanding and believes that the *mantra*, its deity, and its reciter are separate entities.

Great sages who were seers of mantra spoke of two types of mantras: *jada*, or inert, and *caitanya*, or alive.

Japa in the Four Bodies

Vedānta which contains the knowledge of eternal truths, is supreme among scriptures. It describes the four bodies (*sharira*) of the soul (*jīvātman*): the gross (*shola*), the subtle (*sukshma*) the causal (*karana*) and the supracausal (*mahākārana*).

To understand the exact location, size, form, and nature of the different bodies is not at all easy.

This mysterious *Citi*, taking the form of a mantra, resides in the syllables of the mantra.

The Path of Knowledge

Self-realization is not attained without knowledge, for knowledge is the very nature of God. The Supreme Being blazes forth as knowledge. Without knowledge it is not possible to have the final experience. God in fact, is ever present within our inner being.

The Vedantic philosophy of equality is based on true vision. Pleasure and pain, gain and loss, wealth and poverty are all consequences of *Kārma*.

Comments

Knowledge is most important in all endeavors of life, in society, in science, in spirituality. Meaning of knowledge differs of course from field to field, at different levels of existence. Awareness and knowledge are necessary for progress in any activity. Without them, pleasure leads nowhere. As such, real pleasure, the bliss of spirituality, resides in the awareness of the eternal *Brahman*, the one without a second. Whether we call God *Brahman* or Śiva or any other name, the essence of the Universal Self is the same throughout all ages and all traditions. The eternal teachings of the Great Teachers in summary refer to God as "love, delight, supreme happiness, absolute joy, and eternal bliss" [6]. This is the message of all religions, all paths of yoga, all spiritual knowledge and practice.

In the *Bhagavad Gītā*, Lord Krishna explains spirituality, the paths of yoga outlined above. As Lord Buddha also taught, identification with the limited being of body-mind is ultimately the source of all limitations in human life, the suffering that derives from the ego. *Bhagavad Gītā*, the book of the Lord, is an eternal, invaluable practical treatise of the nature of God, disciple and the processes that tie them, what is true *yoga*. This is also the true meaning of complementarity in science, both quantum mechanics and general relativity, wherein, the subject, object and the processes between them provide the true understanding of modern science.

In the section discussing the levels of sound, we describe that there are four levels of sound occurring in the inner subtle levels, the cakras, experienced in deep meditation, the interesting point is that they are connected to the four bodies.

Complementarity and the union of pairs of opposites could not easily be understood if the law of Kārma, past impressions rooted in the subtle realms, was not operating across the universe. Complete path of Yoga in all aspects of spirituality is the path of *Siddha Yoga*, https://www.siddhayoga.org/

Mind

Swāmī Śivananda in his treatise of the Mind [10] discusses the mysteries of the mind and ways to control it.

What is Mind?

The Mind—A Mystery

The vast majority of men know not the existence of the mind and its operations. Even so-called educated persons know very little of the mind subjectively or of its nature and operations. They have only heard of a mind…Mind is nothing but Atma Śakti. It is brain that wants rest (sleep) but not the mind. A Yogi who has controlled the mind never sleeps. He gets pure rest from meditation itself… In Sankhya philosophy, *Mahat* is the term to denote "cosmic mind" or "universal mind" [p. 3-5].

Mind and Body

Thoughts Make the Body

The actions of the mind alone are indeed actions; not so much those of the body.

Face: An Index of the Mind

Mind is the subtle form of the physical body.

Mutual Influence Between Mind and Body

The mind is intimately connected with the body.

Bad Thoughts, the Primary Cause of Disease

The primary cause of diseases which afflict the body is bad thoughts. Whatever you hold in your mind will be produced in the physical body.

The Root of All Evils

The erroneous imagination that you are the body is the root of all evils.

Pain is in Mind Only

Pain is evident so long as you connect yourself with the mind. [p. 2-26]

Mind, *Prāṇa* and *Kuṇḍalinī*

Prāṇa—the Outer Coat of Mind

There are two principal Tattvas in the universe, mind and prāṇa.

Inter-dependence of Mind and Prāṇa

Prāṇa and mind stand to one another in the relationship of the supporter and the supported.

Mind and Kuṇḍalinī

Even a Vedantin can get superconscious state only through the awakening of *Kuṇḍalinī Śakti.*

When Kuṇḍalinī moves from *cakra* to *cakra*, layer after layer of the mind opens up…When Kuṇḍalinī reaches the *sahasrāra cakra*, you are in *Chidākāśa*, (pure) knowledge space. [p. 28-31]

Three *Guṇas*

Sāttva, is purity, favorable for liberation. All *Sādhanās* (spiritual practices) aim at the development of sāttva guna and the attainment of pure, irresistible Will.

Rājas, tendency to look into the defects of others. It also remembers the bad deeds or wrongs done by others and forgets easily their good acts. An ordinary worldly-minded man can hardly hear the inner voice of Atman.

Tāmas, has the characteristics of inertia, darkness, ignorance. [p. 42-44]

The Psychic States

The Two Kinds of Moods and Their Effects

In *Vedānta*, there are only two kinds of moods, *Harsha* (joy, exultation or exhilaration) and Śoka (grief or depression). In the mind these two kinds of moods prevail.

The Meditative Mood

But, there is one good mood in those who practice meditation. It is termed the "meditative mood"…When this mood manifests, you must immediately give up reading, writing, talking, etc.

Imagination

Carefully mark the ways of the mind. It tempts, exaggerates, magnifies, infatuates, unnecessarily alarms through vain imagination, vain fear, vain worries and vain forebodings. [p. 48-50]

Śuddha Manas and Aśuddha Manas

Śuddha Manas or *Sattvic* mind (pure mind) and Aśuddha (impure) Manas or the instinctive mind or desire-mind as it is called are the two kinds of mind according to *Upaniṣadic* teaching. There is the lower mind filled with passion. There is the higher mind filled with Sattva (purity).

Characteristics of the Sattvic Mind

A Sattvic mind likes solitude, silence, simple living, high thinking, philosophical discussions, concentration of mind and company of *Sādhus*, *Mahātmās* and *Sannyasins*.

Characteristics of the *Rajasic* Mind

A Rajasic mind likes crowded cities, much talking, luxurious life, low thinking, the company of women, study of romantic novels, eating dainty dishes and selfish works.

Sattvic Mind Needed for *Ātma-Vichāra*

A sharp, subtle, one pointed Sattvic (pure) mind is needed for *Ātma-Vichāra* (enquiry into *Ātman* or the Supreme Spirit) and study of Upaniṣads. (Note: Vichāra means *deliberation*).

How to Purify the Mind

As one iron shapes another iron, the pure mind of a person which makes efforts in the virtuous path should correct and mould his impure mind. [p. 60-63]

Theory of Perception

Mind Alone Creates Differences

The eyes present before the mind some forms or images. It is the mind that creates good and bad forms.

Mental Cognition Takes Place Serially

Mind can have only one idea at a time…The human mind has the power of attending to only one object at a time, although it is able to pass from one object to another with a marvelous degree of speed, so rapidly in fact, that some have held that it could grasp several things at a time.

How *Brahman* Perceives

In the mind, will and sight are separate. In pure *Cit*, will and seeing are one; will and sigh are combined and no longer, as in the case of mind, separate from each other. [p.70-77]

Citta and Memory

What is Citta?

Subconscious mind is termed Citta in Vedānta. Much of your subconsciousness consists of submerged experiences, memories thrown into the background but recoverable.

The Field of Subconscious Mentation

You repeatedly fail at night to get the solution for a problem in arithmetic or geometry. In the morning, when you wake up, you get a clear answer. This answer comes like a flash from the subconscious mind.

Memory

Smriti or memory is a function of Citta (subconscious mind).

Power of Memory

…When you show symptoms of losing your memory, as you grow old, the first symptom is that you find it difficult to remember the names of persons. The reason is not far to seek. All the names are arbitrary. They are like labels. There are no associations with the names. The mind generally remembers through associations, as the impressions become deep thereby. [p. 79-83]

Saṃskāra

Memory—A Revival of Saṃskāra

The saṃskāras (impressions) are embedded in the subconscious mind or Citta.

Cyclic Causation of Thought and Saṃskāra

An object awakens or revives Saṃskāras in the mind through external stimuli. Hence, a *Sankalpa* or thought arises subjectively from within, without a stimulus from outside.

Karma

The gross body and the mind have, on account of your past Karmas, a tendency to act in a certain way and you act just in accordance with that tendency like a machine.

Your next life will depend very largely upon the Karma you perform in this birth.

How to Acquire Good Saṃskāras

When you repeat *Om* or the *Mahavakya* of the Upaniṣads "*Aham Brahmāsmi*" once, one Saṃskāra of the idea "I am Brahman or the Absolute" is formed in the subconscious mind. The object of doing *Japa* or silent repetition of Om 21,600 times daily is to strengthen this Saṃskāra.

Sankalpa

Sankalpa only is *Saṃsāra*

It is the Sankalpa of the mind that brings about this world with all its moving and fixed creatures. The poisonous tree of the great *Māyā's* illusion flourishes more and more out of the seed of the mind's modifications, full of Sankalpa, in the soil of the variegated enjoyments of the world.

Māyā is a big poisonous tree.

Annihilation of *Sankalpas* Constitutes *Moksha*

It is only Sankalpa of the mind destroyed beyond resurrection that constitutes the immaculate Brahmic seat (Note: Moksha is liberation). [p.94-96]

Thought Creates the World

Thought, the Origin of Everything
Everything in the material universe about us had its origin first in thought. From this, it took its form…The universe is rendered visible through the mind.
Non-Existence of the World—What It Means
The play of mind arising out of *Caitanya* (pure consciousness) constitutes the universe. [p. 99-104]

Avidya and *Ahaṁkāra*

Ahaṁkāra—How It Develops
Atman in conjunction with the *Buddhi* is Ahaṁkāra. The basis of Ahaṁkāra is Buddhi. (Note: Ahaṁkāra is ego; Buddhi is Intellect)… The seed of mind is Ahaṁkāra…This idea of "I" will bring in its train, the idea of time, space and other potencies.
How to Eradicate Ahaṁkāra
The deep roots of Ahaṁkāra should be burnt by the fire of knowledge… You cannot, all at once, eradicate Ahaṁkāra, altogether. [p. 106-108]

The Power of Thought

Thought is a Living Force
Thought is a vital, living force, the most vital, subtle and irresistible force that exists in the universe.
Thought Weaves Destiny
That which man thinks upon in one life, he becomes in another. [p. 110-112]

Vāsanās

Vāsanās—How They Manifest

Vāsanā (desire in subtle form) is a wave in the mind-lake.

Conquest of Ahaṁkāra

If you destroy egoism (Ahaṁkāra, this false little "I") and control of *Indriyas* (senses), the Vāsanās will die by themselves. [p. 132-136] The rest of this great text on the *Mind*, truly a map of the mind in all its aspects, refers to the following additional topics (many already discussed in detail):

How to Control Desires. Pleasure and Pain (pertaining to mind). Real Happiness Lies Within. The Golden Mean (keeping the mind in a state of moderation).

Viveka (wisdom). *Vairāgya* (dispassion, detachment) opens the gate to Divine Wisdom. Intense *Vairāgya* necessary for *Moksha*. Delusion proceeds from Affection.

Renunciation Brings About *Moksha*. Control of *Indriyas*. Fasting. Keep the Mind Ever Engaged. Anger, Prejudice, Intolerance, Insolence, How to Eradicate Hatred. *Moha* (delusion) and Its Cure. Cultivation of Virtues. How to Control the Mind. Mind is the Cause of Bondage and Liberation. Take Everything as It Comes. Destroy the Ego. Maintain a Positive Attitude. Concentration, the Key to Peace. Meditation (discussed in great detail in several chapters). Experiences in Meditation. Rapture or Ecstasy. Meditation and Work. *Samādhi* (intense concentration). *Jñāna Yoga* (Yoga of Knowledge). Mind and *Brahman*. The Dawn of Jñāna. Necessity for a Guru.

Kuṇḍalinī, Cakras, levels of Sound

The sacred knowledge, derived from the (Triadic) Śaivism and the Tantra / Kaula systems, is related to Kuṇḍalinī, and was a hidden knowledge for most people. Yoga and Kuṇḍalinī are intimately connected, both are in the inner realms. Swāmī Vishnudevananda explains, there are four different yoga, however they are one and the same. The only difference is that the mind is involved in different aspects. When the mind dominates the life of the aspirant, it leads to ignorance, Swāmī Vishnudevananda states when "you identify with the body, there is karmic reaction".

Swāmī Lakshamanjoo in the revealed work, *Light on Tantra in Kashmir Shaivism: Abhinavagupta's Tantraloka, Vol. 2* [2] discusses a vast number of issues associated with spiritual path (in the Appendix, which also discusses the *tattvas*). He also presents the attributes of *Śivahood*: "In this highest state of supreme God Consciousness, *anuttara*, there is no need of spiritual progress, no contemplation, no art of expression, no investigation, no meditation, no concentration, no recitation, exertion or practice. Tell me then, what is the supreme and well-ascertained truth? Listen indeed to this! Neither abandon nor accept anything, enjoy everything, remain as you are! In reality, there is no such thing as birth and death, so how can the question arise of bondage for the one who is entirely free, and, therefore, to struggle for liberation is useless and no more than delusion…It is all based on deceitful perception which has no substance. Neither abandon nor accept anything, remain as you are, well established in your own Self. (further down)…all these states with which the universe is formed shine as mutually different but in reality they are not. Whenever you perceive the specificity of some thing, at that very moment you should perceive the essence of your own consciousness as one with it…why not take delight in the fullness of that awareness!"

These are the words of the Master emanating from the teachings of the great Abhinavagupta. All these points lead to the realization of what I call *Illumination*, the Awakening of the supreme Light of Consciousness. One may ask, why the spiritual practices? As all great Teachers declared and declare, to bring the mind, to unite the limited mind into the Universal Mind of Consciousness. This is true Illumination.

The following passages are from [2] and [18]:

Kuṇḍalinī Śakti is the revealing and the concealing energy of Lord *Śiva. On the one hand, it is the revealing energy and on the other hand it is the concealing energy. This* Kuṇḍalinī Śakti is not different from the existence of Lord *Śiva, just as the energy of light and the energy of heat are not separate from the fire itself.* Kuṇḍalinī is *Śiva himself.*

In Trika *Śaivism*, Kuṇḍalinī, which is the internal serpent power existing in the shape of a coil, is divided in three ways:

Parā Kuṇḍalinī functioned by Lord Śiva

Cit Kuṇḍalinī functioned in Consciousness

Prāṇa Kuṇḍalinī functioned in breath

Parā Kuṇḍalinī is supreme Kuṇḍalinī. Is not experienced by *yogins*. It is so vast and universal that the body cannot exist in its presence. It is only experienced at the time of death, it is the heart of Śiva. *This whole universe is created by* parā Kuṇḍalinī, exists in It, gets its life from It, and is consumed in parā Kuṇḍalinī. When this parā Kuṇḍalinī creates the universe, Śiva conceals his real nature and is thrown into the universe, He becomes the universe. There is no Śiva left which is separate from the universe. This is His creative energy. And when Kuṇḍalinī destroys the universe, Śiva's nature is revealed. Therefore the creative energy of the universe is the destructive energy for Śiva and the destructive energy for the universe is the creative energy for Śiva or the concealing energy for the universe and the revealing energy for Lord Śiva. Parā Kuṇḍalinī is the supreme *visarga* of Śiva. Visarga (:) comprises two points, *Śiva and* Śakti or in some sense the revealing point and the concealing point.

Cit Kuṇḍalinī is experienced by *yogins* by means of concentrating on the center between two breaths, thoughts or actions, between the destruction and the creation of any two things. It is happiness and bliss beyond description. In addition to the bliss, you also realize the reality of the Self. You recognize your real nature, you know " I am only bliss, *ānanda*, and consciousness, *cit*". The full rise of Cit Kuṇḍalinī takes place only by the grace of your master and by the grace of your own strength of awareness.

Prāṇa Kuṇḍalinī also comes through the process of centering, It is, however, is only experienced by those *yogins* who along with attachment to spirituality also have attachment to worldly pleasures. If you do not have attachments to worldly pleasures and are only attached to spirituality, then the rise of Kuṇḍalinī takes place in the form of Cit Kuṇḍalinī. There is nothing you can do to determine how the rise of Kuṇḍalinī will take place. It rises in its own way, depending on your attachments.

We note that as in Vedānta which asserts that God is infinite existence, infinite consciousness and infinite bliss; as in Kashmir *Śaivism*, which asserts the triadic nature of Śiva, Śakti and Nara; and numerous other trinities: 1) *Parā* - the highest, supreme 2) *Parāparā* - the middle state consisting of identity within differences, and the last 3) *Aparā* - which is the state of difference, i.e. we and Śiva, "we are separated". Also, consists of the triad of the three fundamental Powers: 1) Will, 2) Knowledge, and 3) Action. Kuṇḍalinī is also triadic, three-natured, Trinity Herself.

The tattvas, the cakras and the rise of the Kuṇḍalinī energy present a seamless body of inner knowledge and experience, valid in ancient times as they are valid today. Therefore, here we present some connections that are very relevant to our spiritual path, with inner revealing experiences. This is what we mean by spiritual qualia (spiritual experiences), giving us the prospect of roaming in the vast inner spaces of Consciousness as we live our daily lives. Specifically, Kuṇḍalinī is the Divine energy which is normally inert, wrapped or "dormant" at the root / base cakra, the *mūlādhāra* cakra (Figure 9). She is coiled 3 ½ times around the Śiva Lingam (sacred symbol of Śiva). Kuṇḍalinī sustains the physical and vital energies but is dormant in the spiritual realm. When awaken She moves the aspirant to the path of Enlightenment.

Each cakra has energy "petals" with a distinct sound represented by a different Sanskrit letter. Also, each cakra has a characteristic *bija* mantra. The 50 letters of the Sanskrit alphabet are each on a different petal, with different numbers of petals in each cakra. Many times in deep meditation Kuṇḍalinī is presented as a light column or a snake (coiled) of Light. The word *kuṇd* means "burning", i.e. Kuṇḍalinī burns the limitations and impressions from the past. Kuṇḍalinī is not hot, on the contrary it gives a warm-cool experience to meditation. It rejuvenates the body and eventually, gradually but steadily, guides the spiritual aspirant to Liberation, with processes based on, and giving indescribable spiritual bliss, Ānanda, to the spiritual seeker.

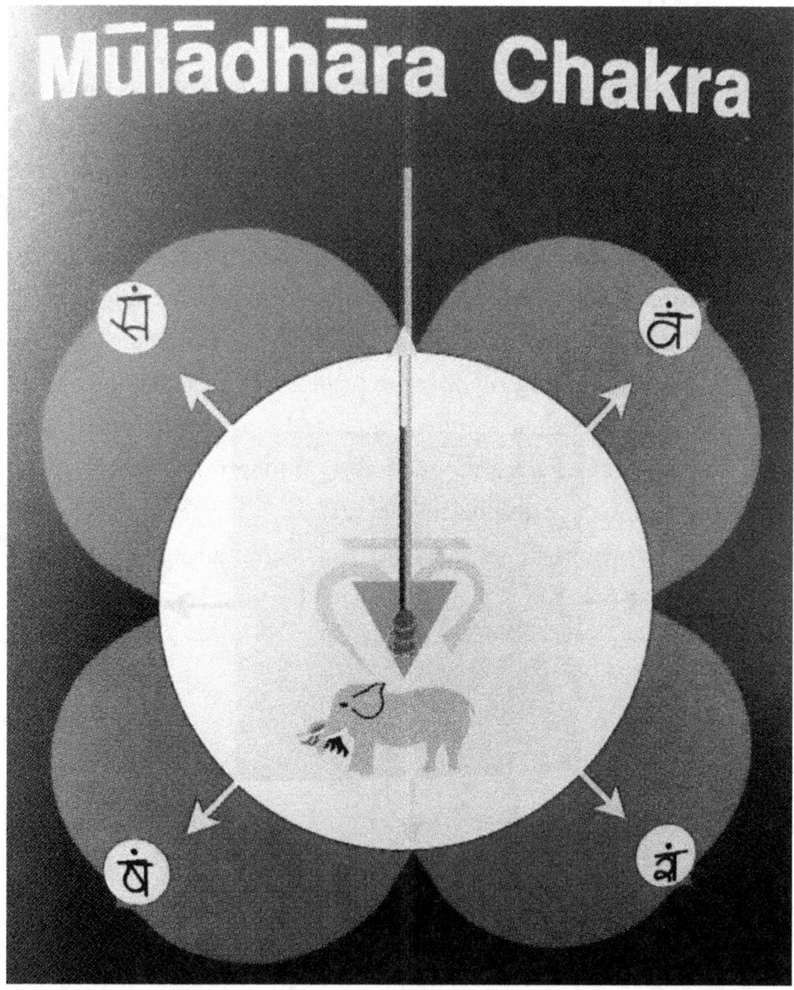

Figure 9

The term cakra means "wheel" or "circle", that is, means of "movement". There are seven basic spiritual light centers of Kuṇḍalinī, where the inner energy channels converge. As Kuṇḍalinī rises, She purifies the various cakras (from primordial negative impressions) and the spiritual disciple becomes more and more in tune with the real Self within.

When Kuṇḍalinī reaches the *anāhata cakra* (figure 10), the inner heart, and "cleanses" it, then there is a balance between the lower trends of being and the higher ones. The spiritual disciple begins to live in the constant experience of Love.

From there on, the inner journey becomes more refined, more subtle, deeper. When the Kuṇḍalinī Energy reaches the ājña *cakra* (the so-called "third eye"), the mind begins to become the universal Mind, with insight, wisdom and inner peace. The individual begins to approach universality, but continues to return to the, now, finer experience of duality, while the inner vision pushes such an aspirant "up" towards the higher levels of *sahasrāra*. The table below (Figure 11) shows 4 of the 7 main cakras. The element *Mahat* at the ājña *cakra* is the primordial, first manifest principle that contains all individual *buddhis* (intellects) and all potential matter of the physical universe. Each cakra contains, among other things, a certain number of petals with corresponding letters, an animal, a sense, an element, a symbol and its own bija mantra, as well as a masculine and feminine deity. The mysteries of the cakras are unfathomable, they are the marks to the inner, vast universe.

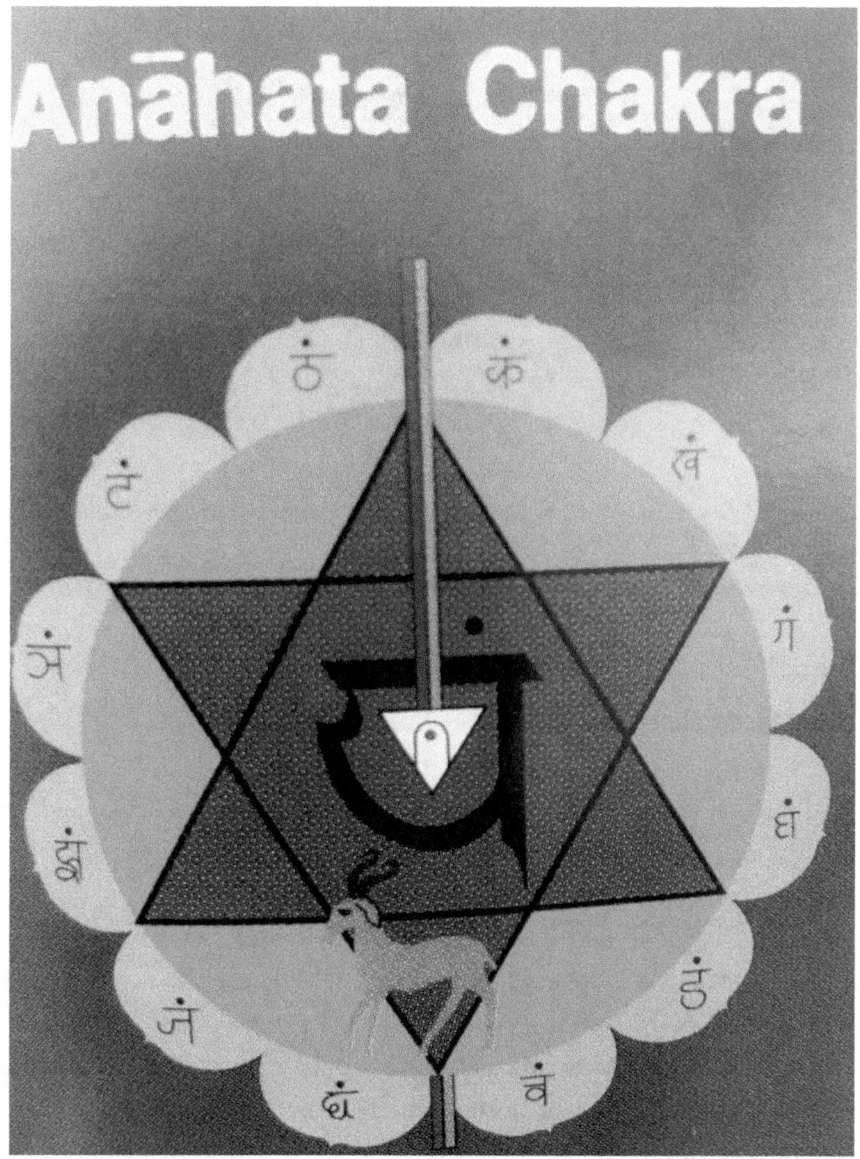

Figure 10

Cakra

	mūlādhāra	*anāhata*	*ājñā*	*sahasrāra*
Number of petals	4	12	2	1,000
Element	Earth	Air	Mahat	
Tanmatra	Smell	Touch	Mind	
Animal	Elephant	Antelope	Swan	
Symbol	Square	Six pointed star	Triangle	
Bija	*Laṁ*	*Yaṁ*	*Oṁ*	

References: *Sat Cakra Nirūpana An Investigation of the Six Cakras*. Composed in the 16th century by Purnananda Swami
David Frawley (2010). *Mantra Yoga and Primal Sound*. Lotus Press, Twin Lakes, WI
Swāmī Kripananda (1995). *The Sacred Power*. Siddha Yoga, South Fallsburg

Figure 11

Levels of Sound

There are four levels of sound that occur in the cakras, in deep meditation, and the interesting thing is that they are connected to the four bodies:

Parāvāk: Silent / open sound, pure Existence, where primordial Consciousness resides. At the root cakra, its element is air. It corresponds to the fourth body, *Turya*.

Paśyantī: Pure feeling, without us realizing anything in particular, the mental sound. It is located from the root to the navel. It corresponds to the third body, to deep sleep. (It is the "point", which contains the whole universe, the well-known *Bindu*).

Madhyamā: Staying in thoughts. From the navel, to the heart cakra, is the inner sound, *Nāda*, sounds heard in meditation. It corresponds to the second body, to dream sleep.

Vaikarī: Normal speech using the neck, tongue, teeth and lips.

Sound Levels

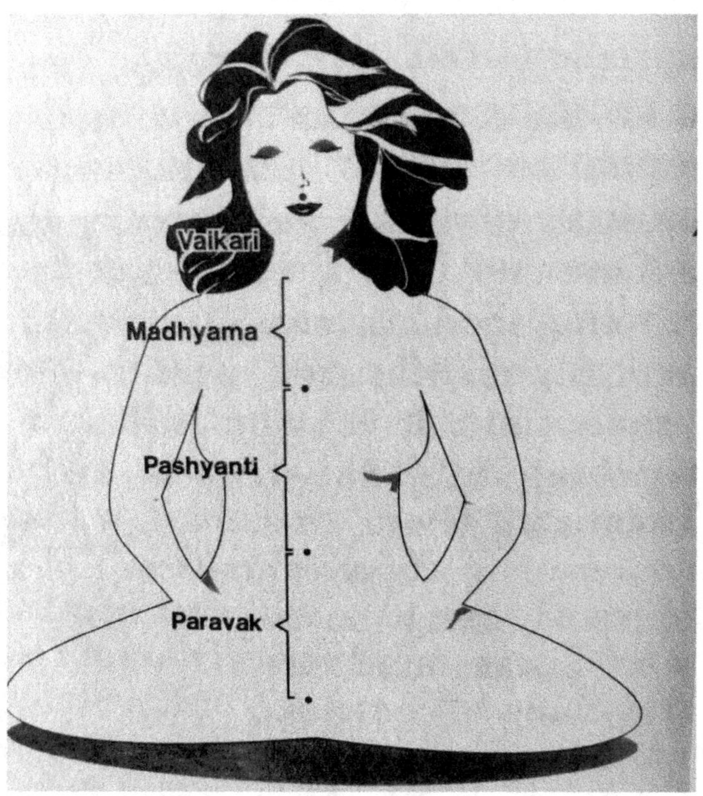

Figure 12

Vijñānabhairava Tantra

Contemporary Śaivite Master Swāmī Lakshmanjoo calls this important tantric text "the practice manual for consciousness of the (real) Self", which is pure Awareness. It consists of 112 meditations or ways of concentrating. They certainly help the rise of Kuṇḍalinī. For example, *cit Kuṇḍalinī* is experienced by concentrating at the center between two breaths, two thoughts or actions, between the disappearance and the creation of any two things. It enters at the junction between any two situations, waking, dream, or deep sleep states. It offers the spiritual disciple ultimate happiness and bliss. It includes spontaneous sudden breathing and upward movement through all the *cakras* up to the *brahmarandhra* opening at the top of the skull. All the worldly gratifications pale compared to this bliss. It consists of 112 meditations or ways of concentrating. The meditations are intimately tied to the inner rise of the fire of Kuṇḍalinī.

Figure 13

Meditation, Mantraḥ Practice and Breathing

Names of Śiva

There are vast numbers of names of Śiva in the cosmos. The word **Śiva** means *Auspicious*. Other names are:

OM, AUM

The Universal, primordial Sound, the *Logos,* the vibration of the universe, the deeper essence of Existence, Consciousness and Bliss. Om is actually made of three sounds expressed as the letters *A, U, M.* The three sounds are universal principles, specifically creation, sustenance and dissolution of the universe as well as all living beings, conscious beings, all objects, all planets, stars and galaxies in the cosmos.

Om Namah Śivaya: Great Liberating *Mantraḥ*

Countless sages, yogis achieved the highest state of Existence, they achieved liberation from the fetters of limited existence, limited understanding of life, by repeating this great Mantraḥ; I bow to Śiva, by repeating the name of Śiva, I recognize Śiva in everything. The *panchakshara mantraḥ*, the five syllable *Namah Śivaya* appeared in the ancient Vedic hymn, *Śri Rudram,* as an adjective where **Śiva** retains its original meaning *auspicious, benign, friendly.*

The five syllable *mantraḥ* together with the primordial sound *Om,* form *Om Namah Śivaya,* the great *mantraḥ* which protects the person who repeats it, and also provides an auspicious way to meditate. It means "Om, adoration (*namas*) to Lord Śiva", where again *Om* is the universal vibration, the original Logos. It also means "Om, I bow to Lord Śiva"; "Om, I offer to Śiva a respectful invocation of His Name". The great five syllable Namah *Śivaya* leads to *Śiva.* This is the deeper meaning of Om Namah Śivaya in Kashmir Śaivism. Practice the great Mantraḥ *Om Namah Śivaya*

Prana

Breath is not just a physical process, it derives from Prana, the very essence of all life, the essence of all yogic practices. Breathing is central to life, when the breath stops, life ends. How to breath, how to focus on the in-going and outgoing breath, how to focus on the points inside and outside the breaths, is also yogic practice. "Prana is not the physical breath, it is the psychic breath. It exists in the nervous system" *Swāmī Vishnudevananda also states "What is this physical breath? What does it have to do with control of your mind?"* And further, "the mind cannot function without prana". Concentrating on the sound of the breath leads one to the exalted prana. Listen to the sound, practice: *Hamsa, Soham.*

More information and practices can be given at many workshop sessions that the seeker may attend.

Important Ideas (Science and Non-Duality)

Therefore, there are local observers who "think" that they are independent and that they are acting. That's how most people think. Our position is to extend the von Neumann / Copenhagen interpretation, which we call "Enhanced Orthodox Interpretation of Quantum Mechanics" and enhances the Copenhagen Orthodox Interpretation. In fact, according to Śaivism, there is only Consciousness (Śiva) and the Power (or Powers) of Consciousness (Śakti) that form all experiences in the universe, making the subject and object appear as different. The two, Śiva and Śakti, are in essence always in union, however, because of Māyā (which is the most difficult power of Śiva to be understood by the simple mind, although the Śaivite Sages insist that it emanates from Śiva himself), who is the universal power of Concealment, they appear as different. Śakti is what in many schools of thought is called *Prakṛti* (Nature), but we believe that the term Śakti offers perhaps less confusion, as Śakti is really Śiva itself, NOT separate or different from Śiva, it just appears like is different. In Śaivism, above the level of Māyā (here we write Māyā with capital M to emphasize its universal role), the so-called *māyā tattva*, there is Śakti, and beneath the māyā level / tattva there is Prakṛti-Nature.

As we saw before, the Tattvas describe the different levels of reality. Traditionally in *Samkhya*, the dualistic orthodox system of Hindu philosophy, there are 36 Tattvas. Abhinavagupta added two additional states *Mahamāyā and Guṇa tattva. Mahamāyā*, the great Illusion, is "above" ordinary *māyā*, the latter being the illusion associated with an individual being. Swāmī Lakshmanjoo [2] explains in detail, in brilliant exposition, Abhinavagupta's timeless rationale: Mahamāyā is the gap between pure subjectivity (Figure 5) and limited subjectivity (Figures 6 and 7). "It is delusion, where you won't know that you are deluded. You will conclude that you are established on truth. But that is not truth, that is not the real thing".

Guṇa tattva is the state where the three *guṇas* first manifest in a state of equilibrium. "In *prakṛti*, you can't see the three *guṇas* because *prakṛti* is the seed state of the three *guṇas*. It is why in Śaivism we have put the element of *guṇa tattva*. Moreover, the three qualities of Nature that are aspects of *guṇa*, namely *sāttva, rājas and tāmas*, provide the constituencies of Nature. As we saw before, *sāttva* (goodness or light) is the quality of purity. The second quality, *rājas*, is the dynamic aspect, and is often associated with the limited mind. The third quality *tāmas* denotes dullness, inertia, ignorance, the "stuff" that is often associated with the limited ego. Abhinavagupta states that initially, in the state of perfection, in the *Past*, they were in perfect balance. However, in the universe of evolution, because of impressions and past Kārmas, one or more of them ended up dominating limited existence and Nature.

Śakti and Śiva are never separate. This is the difference between non-dual Śaivism (the right terms are Trika Śaivism — Triadic Śaivism, or Kashmir Śaivism) and non-dual Vedānta. There are, of course, semi-dual and completely dual versions found in Indian philosophy and other philosophical systems, both in Śaivism and in Vedānta, but many people (who do not know Indian philosophical systems) do not know all these.

Śiva is «passive» and Śakti is "energetic", to put it simply, like heat and fire, which are not really different.

All that we have been discussing in the present monograph, is introducing non-dual Reality as the foundation of the universe (perhaps one of the most advanced idealistic philosophies, and its expansion to philosophical, spiritual and practical, namely non-dual Triadic Kashmir Śaivism) in a new way of synethsis: It is sublime idealistic metaphysics, with implications for everyday life, closely tied to quantum ontology, it can looked at with an expanded mind, playing a central role in understanding both the outer and the inner worlds as complementary realities. However, it is difficult at present, where we seem to only be accepting an "outer world", for humans to accept this vision (even by quantum physicists) of the Unity of Reality. Slowly though we believe that an expanded, integrated, future vision, will be recognized. After all, what are we? As with everything in the universe and countless worlds and being, we are part of this Cosmic Śiva Dance.

To provide some possibilities here, there may a time in the future when the ancient wisdom of Trika, tied to seeking Reality, associated to inner development and experiences, what we may call *spiritual qualia*, can be perhaps tied to new ways to view universal (applying at all levels) tattvas and a future science that will be looking at the "outer" universe of matter and energy as related to inner subtle level, in One Reality. We saw that there is other information in the tattvas of the Triadic Śaivism System, such as the connection of the levels of existence (the 36 tattvas) with the letters of the Sanskrit alphabet (50) and the numerical order that the letters appear at the various levels. As such, numbers as part of *mathematical* structures can be the connection, a connection made to the cakras, where the letters (and associated sounds), the tattvas, and the arithmetic order, are all in relation to each other. Such a possibility, as a vision of the synthesis between spirituality and mathematics, is presented in Figure 14.

Synthesis: Spirituality & Mathematics

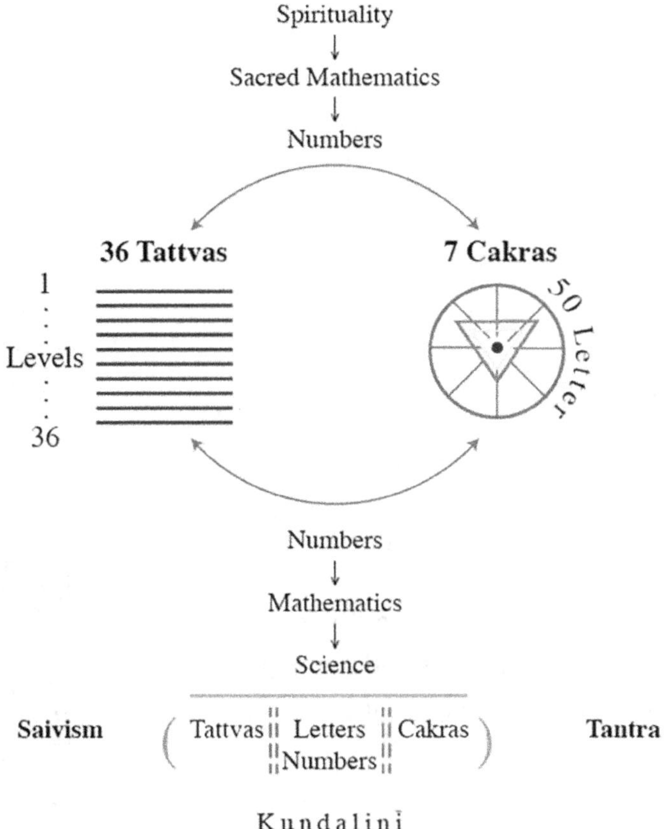

The integral view of mathematics and spirituality shown here is a unified *Holon* and goes beyond the non-dual systems of ancient India. As such, the Śaivism vision is universal, non-dual, idealistic, beyond history and societal structures. It is not surprising that the same vision would apply to the idealism and values of the ancient Greek city states, particularly Athens, the cities of Ionia and *Magna Grecia* in southern Italy. The ancient Greek philosophers Parmenides, Pythagoras, Heraclitus, Socrates, Plato, Aristotle, the Neo-Platonists and Stoics, not only gave us the solid philosophical foundations upon which modern societies and science depend, when coupled to the non-dual traditions of India show us what is highest in humanity. As such, the ancient Platonic and Pythagorean ideas of the mathematical order in the universe reflect the order of supreme spirituality and the future evolution of humanity. The ancient Greek ideas of unity awareness are presented in a simplified way in Figures 15 and 16.

Greek Philosophy - Wisdom

Parmenides
All Is the One

Pythagoras
Highly Structured Way of Life
The School: *semicircle*
5 year waiting period for admission

Heraclitus

- All entities move and nothing remains still
- Everything changes and nothing remains still
- You cannot step twice into the same stream

Figure 15

Greek Philosophy - Wisdom

Socrates

One thing I know is that I don't know anything

Plato

Plato brought philosophical dialogue to perfection The *Academy*

Ideas

Cave & Shadows

Souls

- *Timaeus*
- *Kritias*
- *Politeia*

Figure 16

Notes

For *Śaivism*, see the important work [1]. For what *Śiva* is, and relevant general information, see reference [12]. For the Śiva *Sūtras*, see [3] and [12]. For the *Vijñāna Bhairava* Tantra see [4] and [33] and for meditation methods, especially breathing, see also the same [4]. For the *Pratyabhijñā-hṛdayam* see [9]. For an introduction to *Kuṇḍalinī* see. [2] (There are many relevant books and ancient writings on cakra, eg [19]). For the relationship between science, ancient scriptures and life, see [2], [3] and [10]. The inner experiences arising from the awakened to Kuṇḍalinī Śakti in what truly constitutes Yoga, see Baba Muktananda's autobiography [5].

Pronunciation of Sanskrit Letters

ā	like a in psalm
c	as in chair
ī	like ee in meet
ṛ	like ri in Rita
ṅ	as in sing
ñ	as in new
ṇ	as in Monday
ḥ	visarga is pronounced in the articulating position of the preceding vowel
ś	as in sure
ṣ	as in bush
s	as in sun

References

[1] Swāmī Lakshamanjoo, *Kashmir Shaivism.* Editor. J. Hughes, Universal Shaiva Fellowship (2003).

[2] Swāmī Lakshamanjoo *Light on Tantra in Kashmir Shaivism: Abhinavagupta's Tantraloka, Volume 2, Chapters Two & Three.* Lakshamanjoo Academy Book Series, Hughes Family Trust (2021).

[3] Swāmī Lakshamanjoo, Śiva *Sūtras: The Supreme Awakening.* Universal Shaiva Fellowship (2007).

[4] Swāmī Lakshmanjoo, *Vijñāna Bhairava: The Manual for Self Realization.* Universal Shaiva Fellowship(2008).

[5] Swāmī Muktānanda, *Play of Consciousness.* SYDA Foundation, South Fallsburg, NY (1974).

[6] Swāmī Muktānanda, *Light on the Path.* SYDA Foundation, South Fallsburg, NY (1994).

[7] Swāmī Muktānanda, *Nothing Exists that is not Śiva.* SYDA Foundation, South Fallsburg, NY (1997).

[8] Swāmī Muktānanda, *I Am That: The science of Hamsa from the Vijñānabhairava.* SYDA Foundation, South Fallsburg, NY (1992).

[9] Swāmī Shāntānanda, *The Splendor of Recognition.* Siddha Yoga, South Fallsburg (2003).

[10] Swāmī Śivananda, *MIND Its Mysteries and Control.* The Divine Life Society (1994).

[11] Swāmī Vishnudevananda, *Upadeṣa: Teachings of Swāmī Vishnudevananda.* International Sivananda Yoga Vedānta Centre (2009).

[12] J. Singh, *Śiva Sūtras: The Yoga of Supreme Identity.* Motilal Banarsidass, Delhi (2006).

[13] D.Chopra, M.C. Kafatos, *You Are the Universe.* Random House (2017). And in Greek,

[14] Plato *Timaeus* (in Greek) Πλάτων Τίμαιος, εισαγωγή, μετάφραση, σχόλια Βασίλης Κάλφας. Εκδόσεις ΠΟΛΙΣ (1995).

[15] Aristotle *Nicomachean Ethics.*

[16] M. Kafatos, Th. Kafatou, *Looking in, Seeing out: Consciousness and Cosmos*. Quest Books, Wheaton, IL (1991).

[17] Swāmī Śivananda Radha, *Kuṇḍalinī Yoga for the West*. Timeless Books (1993).

[18] Swāmī Lakshamanjoo, *Kashmir Shaivism, the Secret Supreme*, edit. John Hughes (2003)

[19] *Sat Cakra Nirūpana An Investigation of the Six Cakras*. Composed in the 16th century by Purnananda Swāmī.

[20] D. Chopra, M. K. Καφάτος, *Είσαι το Σύμπαν*. Εκδόσεις Π. Ασημάκης, Μετάφραση Ρ. Καρακατσάνη (2018).

[21] D. Frawley, *Śiva: Ο Κύριος της Γιόγκα*. Εκδόσεις Π. Ασημάκης, Μετάφραση Ρ. Καρακατσάνη (2017).

[22] M. K. Καφάτος, *Βιώνοντας τη Ζωντανή Παρουσία*. Εκδόσεις Μέλισσα, Μετάφραση Σπύρος Πετρουνάκος (2018).

[23] J.C. Chatterji, *Kashmir Śaivism*. SUNY Press, NY (1986).

[24] M.S.G. Dyczkowski, *Spandakārikā* ("The Stanzas of Vibration"). Dilip Kumar Publishers, Varanasi (1994).

[25] M.S.G. Dyczkowski, *The Aphorisms of Śiva*, SUNY Press, Albany (1992).

[26] P.E. Muller-Ortega, *The Triadic Heart of Śiva: Kaula Tantricism of Abhinavagupta in the Non-Dual* Śaivism *of Kashmir*. SUNY Press, Albany, NY (1989).

[27] B.N. Pandit, Aspects of Kashmir Śaivism . Utpal Publications, Santarasa Books, Boulder (1997).

[28] B.N. Pandit, Īśvara *Pratyabhijñā Kārikā* ("Verses on the Recognition of the Lord") of Utpaladeva (translation with commentary). Muktabodha Indological Research Institute, Delhi (2004).

[29] D.B. SenSharma, *Paramārthasāra: The Essence of Supreme Truth of Abhinavagupta (with the commentary of Yogaraja, translation)*. Muktabodha Indological Research Institute, New Delhi (2007).

[30] J. Singh, *Parātrīśikā-Vivaraṇa of Abhinavagupta*. Motilal Banarsidass, Delhi (2000).

[31] J. Singh, *Pratyabhijñā-hṛdayam: The Secret of Self-recognition*. Motilal Banarsidass, Delhi (1980).

[32] J. Singh, *Spandakārikā* ("The Stanzas of Vibration"). Motilal Banarsidass, Delhi (1980).

[33] J. Singh, *Vijñānabhairava* ("Divine Consciousness"). Motilal Banarsidass, Delhi (1979).

[34] *Śri Rudram* Ancient Vedic Chant.

[35] https://www.bhagavadgitausa.com/siva_sutras_all.htm

Triadic Daily Life: Eight Steps plus One

Useful Concepts, Summary

What is the Universe? Who is our real self? The so-called "personality"? Or something other than the body, beyond the human mind and the ego? And what is it? What does modern science say and what does metaphysics, or "Knowledge", say? What are the obstacles to seeing and living the Presence, to know our true *Self*? Basically all spiritual traditions converge on the same message: liberation, or convergence with our essential being, begins with the basic question: "Who am I?"

The answer is: Fundamental Awareness of the Deepest Reality, or Awareness is the deepest absolute reality. Nothing exists without Awareness, which is not subject to the constraints of time and space. The field of full Awareness exists and manifests through our experiences in our daily lives. Experience is understood as the Driving Force of the Universe, that is, according to the quantum mechanics we live in the participatory Universe, Consciousness is fundamental. There are no two worlds, science and everyday life, there is only one world: the quantum. The Three Natural Laws apply to our daily lives, Consciousness views our world as an experience through the three fundamental Natural Laws: *Complementarity* which is the expression of "Yes and / or No", the unity in diversity. *Recursion* (Universality) which is the expression "As here, so everywhere", "In Heaven as on Earth". And *Interactivity*, which is the expression "Everything is process", "relationships and happiness" are directly linked.

Other lessons we learned from quantum physics: There is freedom on the part of the observer and on the side of Nature, true freedom moves the world! Our limits are created by the limitations of observation, by our minds. Time Limits and Mindfulness: How does the quantum world, which is non-local and entangled, look like a classic, a world of complete separation? Here we observe that the role of the mind is paramount, because the covering of the mind conceals underlying Awareness and obscures experience. Understanding natural laws opens up a universe of possibilities in our hands. Then we really understand our minds: Awareness is the source, mind is the tool.

Time and Beyond Time: The inner cave of the Heart, of pure Tranquility. Then we understand the ancient teachings and understand the external illusion of the world and the mind. Fundamental Awareness: Experiencing the inner Awareness in practice.

What is the Universe? Who is our real self? The so-called "personality"? Or something beyond the body, beyond the human mind and the ego? And what is it after all? What does modern science say, and what does metaphysics say, or "Knowledge"? What are the obstacles to see and live the Presence, to get to know our real *Self*? Basically, all spiritual traditions converge to the same message: liberation, or convergence of one with his/her essential Presence, begins by asking the basic question: "Who am I?" If you have doubts and feel anxious, it means do your best but don't expect specific results.

Some of the important concepts we covered in the first and second sections are included in the following charts that are useful to us in the realities of our everyday life, ultimately tying to three fundamental questions about our life, as to where we are going, our here, and who we ultimately are:

Three Natural Laws

- Cosmic Consciousness ***projects out*** world of experience through:

- **Three Natural Laws**
  - **Complementarity (Integrated Polarity)**
  - **Recursion (Universality)**
  - **Creative Interactivity (Flow)**

A shell exhibits the 3 Laws

Five Actions: Cosmic & Individual

- Creation

- Sustenance

- Reabsorption

- Concealment

- Revealing what is concealed

We perform these actions all the time!

Everything in the cosmos does too!

Complementarity ⟷ Intent

Universality ⟷ Knowledge

Interactivity ⟷ Action
The connection is simple! The **key**!
Universe/Nature ⟷ Awareness
Universe/Nature ⟷ Individual
3 Laws ⟷ 3 Powers

- **Where Am I Going?**

- **What is my Purpose?**

- **Who Am I?**

These serve as contemplations to prepare us in a way to look at the 8 practical steps that follow. The mind which, according to the Triadic system, is part of the fundamental Consciousness which is the deepest Reality. The realm of pure Consciousness is present and manifests through our personal experience in everyday life. The 8 rules are the practical steps to be aware of our existence. The Living Presence includes our daily lives. If we consciously experience the Living Presence, with thought and contemplation, with the help of the 8-step practice, our lives will become the abode of eternal happiness.

Eight Steps, Eight Symphonies

Here are eight simple things that we can all observe and do in our everyday lives, a new way perhaps of living, maybe a way to see eight ways as a universal Message. We seem to live our lives with habits, looking for fulfillment outside, seeing differences whether imaginary or "real", blaming others, and what makes us

Symphonies of life

[1] *Expect the unexpected.* Be prepared to make the plans that need to be made. Do not go for the results but make plans to do what is needed to be done anyway.

[2] *Pay attention to little things in our life.* The smaller they are, the more important, the more significant.

[3] *We only see results afterwards, after the specific facts have occurred.* Something small happening unexpectedly, or by synchronicity.

[4] *Tell our mind and the ego to wait!* We can say: Stop, wait. We want to know now and we always want to know in advance.

[5] *One step at a time.* One may ask: Then how do we make progress? If we find *afterwards*, what are we going to do *now*? We don't know now; so we will find out afterwards!

[6] *Things will start moving faster and faster.* When we start following the first five ways, we will see that what might occur later in the future, it will actually happen now, or sooner.

[7] *Remember our own mortality.* First of all, among the little things that we have to pay attention to, is to watch all around us how life and death come and go all the time.

[8] *Things that happen in our life, which give us "hard time", are the ones that really advance us.* The hard things in life, if we look at them the right way, maybe they will teach us.

[9] (the 9th symphony) *Be! We Are!*

This work is both theoretical and practical. It provides us a whole new way of looking at ourselves and with the 8 steps to move forward. Think about what you read, the practices that come from science, the non-duality of the Triadic System, and your daily life. Not because I say it but, hopefully, because it will provide you with useful and lasting tools. Try it, make it all part of your life. One step at a time. Let's follow the Presence which is none other than our *Being* is ours in the adventure called our life, Living with other people we know, we meet. Let this book be a handbook of conscious life, our existence, the Reality we truly are!

Contemplations of our Role as Humans, the Lessons of Life

We thus have at our disposal the basic theoretical elements from science and philosophy. The 8 symphonies give the practical means of *Living* a full life, right from our own homes, from our places of work, moving about, while we communicate with each other in person, in the company of others, or via teleconferencing, telephones, e-mails, texting and all the social media. The 9th symphony is us, our living presence. This Presence all around us is directly apparent when we slow down and pay attention to our being, the eternal Now. The Presence is not inert, on the contrary it is completely alive. It is the conscious Awareness that exists within us all.

As we observe the ecological challenges, the pollution, seemingly reduced, not running around as much, perhaps not consuming as much, seemingly not having as many heart attacks, while the global health system is under stress, the local hospitals running full stop and living with the spreading of fear, perhaps allowing ourselves to take some deep breaths, and the Earth to nurture all species, including us, now some concluding comments and questions to all of us:

- What is the "message" of recent types of health and other challenging issues affecting humanity? What is the message really telling us?
- The 1st symphony is "expect the unexpected". The 2nd is "pay attention to the little things in our life". The 7th is "remember our own mortality".

The 8th, the "great difficulties" that make us evolve. What does all this tell us about situations of great societal challenges?

- Is it a coincidence that a particular virus arrived in late 2019 or early 2020? Was it expected? Was it unexpected?

- Is it part of the natural ecosystem of life? If so, what does it tell us? If it is not part of the natural ecosystem, still the question remains, what does it tell us?

Here are some thoughts and feelings: Could it be that any situation where fear takes over is pointing us to our inner world? Maybe as captives of external circumstances, we can contemplate what we are doing first to ourselves, by the way we have been living, to our society and to the ecosystem, to the entire environment? Maybe we can change how we consume, how we cause pollution and stress?

Maybe instead of *fear*, can we follow *love*? Assessing where are we going in our lives, our search, as an emerging *Noosphere* that Theillard De Chardin talked about? Maybe a *Message* is emerging, pointing to ourselves, our loved ones, our fellow humans, to see all in Unity that we are? To assist each other? To emerge from looking for solutions "out there" to contemplating that life is teaching us something "in here"?

They say we live in the times where divisions, strife and blaming each other rule supreme, where we often threaten each other, even with death and war. Beyond all the threats, all the fear, Love alone, the eternal Light of the Self shines. Is it possible that "situations" like any pandemic, are pointing to the end of the past way of living, to a new way of life, a vision where we take care of each other, where we take care of the ecosystem, the environment? Maybe to a truly new era, the dawn of the New Age, the age of Truth. How exciting that would be! And now going back full circle, the three big questions for all of us to contemplate:

- **Where are we going?**
- **What is our purpose?**
- **Who are we?**

These fundamental questions are for each one of us to ask and contemplate.

References

[1] Swāmī Lakshamanjoo, *Kashmir Shaivism*. Editor. J. Hughes, Universal Shaiva Fellowship (2003).

[2] Swāmī *Sivananda, MIND Its Mysteries and Control. The Divine Life Society (1994)*.

[3] W. James, *The Principles of Psychology*. New York (1890).

[4] Μηνάς Κ. Καφάτος, *Βιώνοντας τη Ζωντανή* Παρουσία, Εκδόσεις Μέλισσα, Μετάφραση Σπύρος Πετρουνάκος (2018).

[5] M. Kafatos, Th. Kafatou, *Looking in, Seeing out: Consciousness and Cosmos*. Quest Books, Wheaton, IL (1991).

[6] M. Kafatos, R. Nadeau, *The Conscious Universe: Parts and Wholes in Physical Reality*. NY: Springer-Verlag (2000).

International Publications

Auroville Architecture
by Franz Fassbender

Auroville Form Style and Design
by Franz Fassbender

Landscapes and Gardens of Auroville
by Franz Fassbender

Inauguration of Auroville
by Franz Fassbender

Auroville in a Nutshell
by Tim Wrey

Death doesn't exist
The Mother on Death, Sri Aurobindo on Rebirth *Compiled by Franz Fassbender*

Divine Love
Compiled by Franz Fassbender

Five Dream
by Sri Aurobindo

Vision
Compiled by Franz Fassbender

Passage to More than India
by Dick Batstone

The Mother on Japan
Compiled by Franz Fassbender

Children of Change: A Spiritual Pilgrimage
by Amrit (Howard Shoji Iriyama)

Memories of Auroville - told by early Aurovilians
by Janet Feran

Featured Titles

Divine Love

The texts presented in this book are selected from the Mother and Sri Aurobindo.

"Awakened to the meaning of my heart. That to feel love and oneness is to live. And this the magic of our golden change, is all the truth I know or seek, O sage."

Sri Aurobindo, Savitri, Book XII, Epilog

A Vision by the Mother

On 28th May 1958, the Mother recounted a vision she once had of a wonderful Being of Love and Consciousness, emanated from the Supreme Origin and projected directly into the Inconscient so that the creation would gradually awaken to the Supramental Consciousness. The Mother's account of this vision was brought out a first time in November 1906, in the Revue Cosmique, a monthly review published in Paris.

A Dream – Aims and Ideals of Auroville
the Mother on Auroville

50 years of Auroville from 28.02.1968 - 28.02.2018
Today, information about Auroville is abundant. Many people try to make meaning out of Auroville – about its conception, to what direction should we grow towards, and, what are we doing here?

But what was Mother's original Dream and what was her Vision for Auroville back then?

Matrimandir Talks by the Mother

This book presents most of Mother's Matrimandir talks, including how she conceived the idea for this special concentration and meditation building in Auroville.

Memories of Auroville - Told by early Aurovilians

Memories of Auroville is a book about the very early days of Auroville based on interviews made in 1997 with Aurovilians who lived here between 1968 and 1973. The interviews presented in this book are part of a history program for newcomers that I had created with my friend, Philip Melville in 1997. The plan was to divide Auroville's history into different eras and then interview Aurovilians according to their area of knowledge. Our first section would cover the years from 1968 till 1973 when the Mother was still in her physical body.

The Way of the Sunlit Path

May The Way of the Sunlit Path be a convenient guide for activating this ancient truth as a support for a Conscious Evolution.
May it illumine the transformation offered to us in the Integral Yoga.

A Dream Takes Shape (in English, French, Hindi)

A comprehensive brochure on the international township of Auroville in, ranging from its Charter and "Why Auroville?" to the plan of the township, the central Matrimandir, the national pavilions and residences, to working groups, the economy, making visits, how to join, its relationship to the Sri Aurobindo Ashram, and its key role in the future of the world. This brochure endeavours to highlight how The Mother envisioned Auroville from its inception, some of the major achievements realised over the years, and some of the difficulties currently faced in implementing the guidelines which she gave.

Mother on Japan

I had everything to learn in Japan. For four years, from an artistic point of view, I lived from wonder to wonder. And everything in this city, in this country, from beginning to end, gives you the impression of impermanence, of the unexpected, the exceptional... ...everything in this city, in this country, from beginning to end, gives you the impression of impermanence, of the unexpected, the exceptional. You always come to things you did not expect; you want to find them again and they are lost – they have made something else which is equally charming.

Auroville Reflected

On 28 February 1968, on an impoverished plateau on the Coromandel Coast of South India, about 4,000 people from around the world gathered for a most unusual inauguration. Handfuls of soil from the countries of the world were mixed together as a symbol of human unity. Why did Indira Gandhi, the erstwhile Prime Minister of India, support this development for "a city the earth needs?" Why did UNESCO endorse this project? Why does the Dalai Lama continue to be involved in the project? What led anthropologist Margaret Mead to insist that records must be kept of its progress? Why did both historian William Irwin Thompson and United Nations representative Robert Muller note that this social experiment may be a breakthrough for humanity even as critics commented, "it is an impossible dream"?

A House For the Third Millennium

Essays on Matrimandir

Nightwatch at the Matrimandir...
A cosmic spectacle; the black expanse above, the big black crater of Matrimandir's excavation carved deep into the soil. The four pillars - two of which are completed and the other two nearing completion - are four huge ships coming together from the four corners of the earth to meet at this pro propitious spot...

Passage to More than India

This book is a voyage of discovery. In 1959 the author, Dick Batstone, a classically educated bookseller in England, with a Christian background, comes across a life of the great Indian polymath Sri Aurobindo, though a series of apparently fortuitous circumstances. A meeting in Durham, England, leads him to a determination to get to the Sri Aurobindo Ashram in Pondicherry, a former French territory south of Madras.

www.ingramcontent.com/pod-product-compliance
Lightning Source LLC
LaVergne TN
LVHW020345200726
843507LV00012B/2506